Fly-Fishing

Practices

A series edited by Margret Grebowicz

Fly-Fishing by Christopher Schaberg
Juggling by Stewart Lawrence Sinclair
Raving by McKenzie Wark
Running by Lindsey A. Freeman

Fly-Fishing

Christopher Schaberg

DUKE UNIVERSITY PRESS
Durham and London
2023

© 2023 DUKE UNIVERSITY PRESS
All rights reserved
Designed by A. Mattson Gallagher
Project Editor: Annie Lubinsky
Typeset in Untitled Serif and General Sans
by Copperline Book Services

Greg Keeler's sonnet "A Glimpse" is used by permission of the author.

Library of Congress Cataloging-in-Publication Data
Names: Schaberg, Christopher, author.
Title: Fly-fishing / Christopher Schaberg.
Other titles: Practices.
Description: Durham : Duke University Press, 2023. | Series: Practices | Includes bibliographical references.
Identifiers: LCCN 2022028087 (print)
LCCN 2022028088 (ebook)
ISBN 9781478019367 (paperback)
ISBN 9781478016724 (hardcover)
ISBN 9781478023999 (ebook)
Subjects: LCSH: Schaberg, Christopher. | Fly fishing. |
BISAC: SPORTS & RECREATION / Fishing | NATURE / Environmental Conservation & Protection
Classification: LCC SH456 .S283 2023 (print) | LCC SH456 (ebook) |
DDC 799.12/4—dc23/eng/20220822
LC record available at https://lccn.loc.gov/2022028087
LC ebook record available at https://lccn.loc.gov/2022028088

Cover text handwritten by Christopher Schaberg.

A Glimpse

I fish the faster water where it slows,
and kick up rocks to chum the pools with nymphs.
Why am I here? The river only knows,
but I'll keep at it till I've had a glimpse
of a brook trout's orange and green, a rainbow's pink
and dipped the colors briefly in the air
of pine and lupine, before I let them sink
back into the shadows that they were.
I apologize to those of you
who haven't felt the river's push and pull
or drunk its air until the mind turns blue
and think that fishing's just a bunch of bull.
I concede, it is a waste of time,
this ploy to sense another through a line.

—Greg Keeler

CONTENTS

Fly-Fishing 1

Afterword 97
Minor-Fishing Lessons

Bibliography 103

Fly-Fishing

I'M STANDING IN WATER UP TO MY WAIST. It's cold, the ice having melted only a few weeks prior. But I'm wearing fleece pants beneath my waders, so I am warm and dry. I'm making my way slowly toward a stand of submerged dogwood trees that are poking up through the surface in about four feet of water. Once I'm about fifty feet away from the red creaturely fingers of the branches, I'll pull a dozen coils of line off my reel, piling it in front of me in the water, and then cast a chartreuse-and-white minnow-pattern fly past the submerged tree, retrieving the fly with darts and twitches to give it a simulated fleeing appearance. I'll bring my fly past each side of the flooded limbs, avoiding them while trying to attract the fish that I know are lying in wait. I can't see what is happening under the water—but I can feel it, imagine it. I'm *almost* there.

I wrote the majority of this book over one summer in northern Michigan, fly-fishing in the inland lakes and ponds that pepper the coastline of Lake Michigan, where I'm from and where I return for a portion of each year. I wrote before I went fishing, after returning from fishing, and even during fishing (in my head, at least, and one time I even audio-recorded myself trying to "write" a part of the book out loud to myself as I fished—that was awkward, and I didn't attempt it again). While I wrote this book, I read various fly-fishing memoirs and guidebooks, and I paid attention to the ways people talked and wrote about fly-fishing, and how they mythologized it. I've attempted here to write my own fly-fishing book. But the truth is that this practice is shared, even as it's highly personal and individualized. And part of the shared aspect of fly-fishing is that it includes more than the people who fish—more than the humans who do it. It always exceeds the tactical mind. I try to reflect on fly-fishing, or project it out onto a map or grid of knowledge, or experience it fully . . . but it always escapes.

Fly-fishing gets its name from the typically small lures that are used, called *flies*. Contra their name, they don't always represent insects. Plenty of them *do* look like bugs: ants, mosquitoes, mayflies, grasshoppers, dragonflies, and all their larval forms. But they can also mimic shrimp, crawfish, frogs, leeches, or small fish. Even mice. In fact, sometimes they don't represent anything at all. I thought it would be useful to begin by describing what flies are, since they are the literal point of attachment between me and the fish that I pursue. A lot of the flies I use look basically like this:

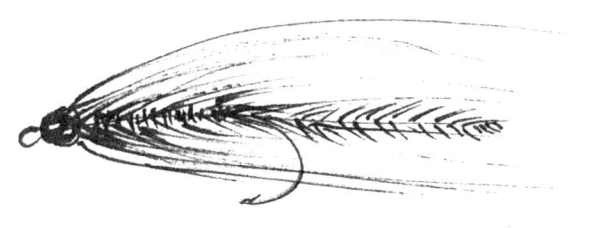

There might be a little red material near the head, small reflective eyes, or some silvery or sparkly strands tied into the body. A fly like this looks stiff and alien when out of water, but when it swims, it comes to life.

I didn't always tie flies—for a long time I felt that it was a waste of time and too much extra crafty stuff—but I do it these days to amplify the experience of fly-fishing. For there is a special thrill that comes with catching a fish using the cumbersome technique of fly-fishing *and* using a fly made by hand.

When I was eleven years old, I fell headlong into an obsession with fishing. I learned how to fish with the intensity of a budding (if somewhat antisocial) adolescent. I devoured books and magazines on the subject, and I absorbed the ephemeral lore of random old-timers I crossed paths with, who would share a productive technique, secret hole, or effective lure pattern. I have a hazy recollection of saving babysitting money, buying an inexpensive fly rod combo kit (a two-piecer, line already on the reel), and practicing casting across the lawn. I don't think I ever used it on the water, and I'm not sure where it ended up.

Nobody in my immediate family fished, but my parents gladly indulged this obsession—probably because it kept me out of the house for hours at a time, pursuing the sunfish and bass that populated the suburban ponds near our home in southern Michigan. I became so engrossed with fishing that I convinced my parents to homeschool me the following year. My mother responsibly researched and purchased a curriculum, but in truth my weekday education rapidly became one-third book learning, two-thirds solitary fishing. But I was learning other lessons, too, as I explored those ponds. I was learning about ecosystems and watersheds, weather and seasons, predators and prey. I was encountering life cycles and fertilizer runoff, die-offs and different species' fragility and resilience, private property and shared commons.

I would occasionally leave fishing for a year or more at a time, but then always return to it with passion. I didn't fish a single day while I was in college, but after I graduated and was working a summer job in Wyoming guiding river trips, I discovered fly-fishing on the streams that coursed through Yellowstone National Park—I was hooked once again. I spent the following two years earning my master's degree while fishing the waters around Bozeman, Montana. Then my rod and flies sat dormant again for six years as I worked on my PhD in Davis, California.

When I moved to New Orleans for a teaching job at Loyola University, I didn't plan on taking up fishing again, even though the license plates bragged about Louisiana being "Sportsman's Paradise." But after a couple years I discovered I could fly-fish a postapocalyptic swoop of the Mississippi River, and then later

I ventured into the lagoons of City Park and around Bayou St. John—not for the biggest fish or the most picturesque scenes, but because I had to do it once I saw fish swimming there. I don't know how to get by, otherwise.

When I can't fish, I try to find time to tie my own flies at home: small puzzles of various materials that, when presented in the right way in the right spot, might just catch fish. Fly-fishing for the small fish near my home in the neighborhood of Mid-City, and catching these fish on the flies I tie, has become a centering—or really, a humbling—activity for me during the school year. And in the summers I fish as much as possible when I'm back up in Michigan.

I'm not going to turn to rods or casting next. In fact, I'm not going to talk about line or knots or gear in any real detail. There are plenty of good and instructive resources that outline just how to tie flies; how to assemble the rod; how to load the reel with backing, line, and leader and tippet; how to tie different knots; then how to cast in different conditions on different kinds of water. But these granular topics, when written down or otherwise recorded, can easily feel too abstruse for the uninitiated—or too academic, for someone who feels fishing in their bones.

I'm taking a different tack. Here I reflect on how fly-fishing has been a constant in my life, and about the lessons I draw from fly-fishing. Or just as often, how I run into conundrums around fly-fishing—while I'm fishing, or preparing to fish, or reflecting on it later. If fly-fishing has become a mainstay in my life, it always leads to farther-flung realms of thought, observation, and practice. The things we do that we love most, the activities that give us peace, or focus, or thrills, are entangled with all the other parts of life. I hope to show how those inevitable entanglements play out around fly-fishing, for me.

Two days ago, here in northern Michigan near the Sleeping Bear Dunes National Lakeshore, I was sitting where I am sitting now as I write this. Two days ago, I was feeling a buildup of anticipation to go fishing the following morning.

I had spent several days preparing for this outing: getting my flies organized and repaired, arranging my leaders and tippets in my vest pockets for easy access, deciding what layers to wear (or to have as backups), and putting my waders and boots in the back of my car. I was literally counting down the hours until I would sneak out of the house at 4:45 a.m. (so as not to wake my three sleeping children) and drive to the spot where I would hike in to a small lake in the nearby national park. I was debating the different possible shorelines where I might fish, imagining the various contingencies and varying conditions that would make me choose one area over the other, or fish on the surface versus underwater.

I spent a lot of time that day mentally planning and visualizing the morning to come. But now I am sitting here again; that morning has passed. Now I'm writing about fishing.

It was a good fishing morning: misty at dawn, mirror-calm water, and fish dimpling the surface of the lake. A pair of nighthawks were looping and diving right at the water's edge. My friend Glen and I caught dozens of fish in the first couple hours, releasing them all back into the water. The sun came up strong in a blue sky and created intense glare on the lake—and Glen had left his sunglasses in his car. This made it difficult for him to see where his fly was landing, or where fish were holding. The midmorning hours were slow. We trudged along in waist-deep water, casting to promising-looking spots where we knew fish should be, but they were skittish. We saw plenty of fish swimming in the shallows, sometimes darting right between us. But the water was so clear and the day so bright that it was, well, clear that they were not easily tricked into biting our flies. The wind also came up and created just enough chop on the water to make precise casting much more challenging.

Toward the end of the morning, we waded over to a protected shallow bay and managed to catch a few more nice fish that were hanging out in nooks between crowded reed stands. We hiked back to our cars tired and sore but satisfied—if also tacitly frustrated that the day had been less than ideal for fishing. It was illogical, really: we probably caught upward of a hundred fish, hadn't seen any other people, had been surrounded by loons and swans and sandhill cranes and green herons and redwing blackbirds—for all intents and purposes, a picture-perfect fishing morning.

But it *could* have been better. It could have been like *today*. Today, as I reflect on yesterday, it has been dead calm all day and mostly overcast with a slow-moving but nonthreatening cloud cover. Now *this* is when the fish are really active. Visibility is better for the fish (for spotting prey), and it is easier for us humans to blend in *and* see the fish moving below the surface. And so much easier to cast. Fly casting can be done in all sorts of conditions, but there's nothing like setting up a perfect cast and seeing it through without the line getting blasted into oblivion by a fierce gust. Today the conditions are perfect.

Instead of fishing, though, today I am writing about fishing. This summer I have been alternating between reading about fishing and actually fishing. And now I am trying to write about fishing. What's so weird is that fishing has always been something I do when I'm *not* writing, or even something I do so I don't *have* to write (or read). So this blending that I'm doing here is a little uncomfortable and strange. I'm also intimidated by the volume of other texts that have already been written on fly-fishing. Just this past winter, as I was starting to outline this book, I was delighted (if also a bit anxious) to see that Mark Kurlansky had a new book out called *The Unreasonable Virtue of Fly Fishing*. I read the book, realizing that our experiences of fly-fishing were very different: his was rather narrowly focused on trout and salmon fishing in rivers (with some brief forays into exotic waters for other fish), whereas my own fly-fishing takes place almost exclusively on a series of lakes near where I grew up, in northern Michigan, and I primarily fish for panfish, bass, and pike. (This is not the standard image of fly-fishing, though it has its own long history.) But I am getting ahead of myself. Or rather, I never really knew where to start in the first place, so I am just jumping in wherever I can. Or whenever I can.

Fly-fishing has something to do with *time*, with the experience of losing track of time but also the anticipation of a time to come—a time that will then recede into the past, and even the highest-definition photographs cannot bring one back to the past present of being there. Losing a sense of time while fishing is a common feeling. As Kurlansky aptly describes it in his book, "There is always that moment, boots on, stepping down into the river, like slipping through a magic portal" (243). And while this is hardly exclusive to fly-fishing, perhaps there is something about the rhythm of fly casting, or the fact that one is more apt to be *in* the water (a physical marker of time, flowing or lapping, even when standing still) when fly-fishing, that makes temporality a distinctly slippery thing here. The *timing* of fishing is also crucial: an hour or even a few minutes can make all the difference. Simply *finding* the time to fish is yet another temporal dimension.

I find myself looking back at my life to limn out when and why and how I started fishing. And I find myself contemplating the role that fishing plays in my present (ongoing) life now. I wasn't sure, at first, whether to call this book *Fishing* or *Fly-Fishing*, because I didn't always fly-fish. I started out as an eleven-year-old kid fishing with worms beneath a red-and-white plastic bobber, and I gradually learned all sorts of artificial lures and how to use different types of reels and rods . . . and only later, after college, did I really take up fly-fishing in earnest. I'll tell that story later.

There are roughly four genres of fly-fishing books: memoirs, travelogues, how-to guides, and histories. Most books on the topic blend or borrow across these different modes. I find myself falling into these modes from time to time, yet resisting them when I recognize them—always seeking to climb out of the ruts of fly-fishing writing conventions.

I'm attempting in these pages to retrace my life in fishing, and also to make connections with ideas and encounters that fishing has made possible for me. Thomas McGuane sums up the genre of the fishing memoir this way: most writers follow a predictable path, "remembering their first waters, their mentors, their graduation through various methods; there is for each of us a need to understand and often to tell our own story in fishing" (252). This description feels accurate to my own experience writing about my life in fishing. At the same time, I'm trying to resist some patterns and expected turns.

My fishing these days takes place predominantly with a fly rod, which is why the book is framed around this specific kind of fishing. But I won't expound too much on what makes fly-fishing special or unique, or the most artistic kind of fishing. Fishing is more instinctive and expansive, both for me and around the world, wherever and whenever it happens. I've ended up fly-fishing owing to a series of random encounters across my life—that's all. I've stuck with it because there are certain (or uncertain) ineffable things I like about it. It's not exactly minimalist, but I have embraced a relatively simple version of fly-fishing.

In about two weeks I am scheduled to take our ninety-six-year-old neighbor Ted on what he described as his "last fishing trip." He has been fly-fishing on the lakes in our region as long as I have but has not been out much over the past five years. (His last time fishing was a guided trip in Montana two years ago, the summer before the COVID-19 pandemic.) When I was talking to Ted on the phone to plan our outing, he ended the call by telling me he was "full of great anticipation." He sounded giddy, like an almost century-old child.

I know that feeling—I have it every time before fishing. How can it be that I get more excited for each fishing trip? Even to the same lake, doing the same thing, trying to catch basically the same fish . . . my excitement mounts in the hours leading up to my outing. It has to do with a strange combination of knowing what to expect—indeed, planning for various possibilities—and *not* knowing what will happen. At the intersection of careful preparation and radical contingency, fly-fishing occurs.

In some ways, it's this feeling that the practice is conditioned on. For some, it may be channeled into obsessive gear curation or elaborate logistical planning. But even around those activities there must be a similar aura of the *thought* of fishing to come. A feeling of great anticipation.

As I've been reading fishing memoirs, an interesting part of this archive is running across lines that feel uncannily familiar—like exact thoughts *I* have had while fishing (or reflecting on fishing), too. But other times, I read sentences that turn me off or disgust me—not my kind of fishing, or not my kind of thinking about fishing. Sometimes it's brash anthropocentrism, or condescension toward fish or habitats. Or it's machismo. Or it can just be a subtle tone that repels me.

I was thinking about what fishing means to me yesterday morning, as I fished. I had woken up at 4:00—too early to leave just yet. I lay there in bed warm and cozy, calculating the appropriate time to creep out of the house and head to the lake. The lake I was headed to was just six miles away, but driving in the predawn darkness can be dangerous, so it takes a little longer than in the full daylight. On my way to the lake, I would see a massive snapping turtle paused in the middle of the road, head held high perhaps to alert vehicles to its presence; so I braked and drove around it. It was on its way back from laying its eggs, most likely. I also saw several pairs of deer eyes in the brush at the edge of the road. And raccoons scurrying off into the underbrush.

But back in bed, I hadn't left yet. I listened to the crickets and the first robin songs, and finally checked my phone and saw 4:31 and quietly pulled on my clothes and slipped out of the house. After driving into the woods and down a dirt road for the last mile, I parked the car and put on my waders and wading boots. I have a simple vest with just a few pockets holding my pliers and flies and some other supplies. The sky is still gray blue, not yet 5:00.

On this lake, which I have been fishing since I was a teenager, I know what I am doing—or rather, I know what I am after. There are not a lot of uncertainties, at least when it comes to preparation and equipment. No, the uncertainties come in the form of all the other elements and life forms teeming around me. I know the trail; I know the shoreline (though each year it shifts, depending on the water level from winter precipitation). I wade in and recognize old stands of reeds, dogwoods growing in encroaching shallows, and clusters of water lilies and pondweed. The water is still and dimpled with rises of fish taking caddis flies from the surface. I cast a dark fly out toward the edge of a wall of reeds. What I'm using is not technically a fly but a *popper*: called that because its floating body has a concave front that burbles and bloops when it is twitched. It also has rubber legs and a delicate feather tail that all wiggle around and behind it. It could look like a frog, or a dragonfly that alighted and became stuck on the surface. Almost immediately, in much less time than it took to compose these sentences, the water opens up and the popper is taken under, and my rod is bent against a strong, chunky bass who is pulling toward the weeds near the shore. I strain to pull it away from the weeds, because if it gets into the resolute stems of the lily pads, it is highly likely that it will leverage the hook out. (Bass have bony jaws with a thin connective membrane that the hook typically goes through; it heals over quickly and does not bleed.) I'm pulling line quickly to bring the fish toward me, away from the weeds—but the bass changes its tactic and instead races toward me. One of the hazards of wading, especially in a still lake, is that a fish will mistake (or outright utilize) your legs for further structure to be leveraged against the line, the fish circling rapidly around your legs and getting *you* tangled up in the mess. This can cause a terrifying confusion and even trip you up when the fish is closest to you, its powerful tail fins churning the water and adding to the chaos. So I am stripping the line in even faster now, trying to keep up with the advancing fish.

As I write this, I feel two conflicting anxieties. On the one hand, I am aware that I am describing details and actions that may be mysterious and hard to imagine (or possibly even be repulsive!) for the reader who has not fly-fished. So I am attempting to carefully explain each part of what happens. On the other hand, I am aware that readers who fish may find these descriptions almost tedious and banal, so common are the experiences to fishing. I am describing fishing scenes. But therein, I suppose, may lie the appeal, for readers who fish: you can *almost* relive the scenes in your mind, and they may trigger recollections of similar moments from one's own outings. Or generate excitement for adventures to come.

My fishing adventures, though, can hardly be called that. They are so familiar. I return to a handful of lakes I adore, places I know intimately. When I was fourteen, I drew a meticulous map of one of the lakes in the national park that I fell in love with when my family first moved to the region. I fished this lake innumerable evenings of my early teen years. My parents would drop me off after school or after dinner, and either I'd hike out to a little point that had a sandy beach, from which I could cast toward a slight drop-off, or I had a flimsy little rubber boat with plastic oars, a foot pump, and a tiny electric motor that ran on a pint-size twelve-volt battery. I would putter across the lake to my favorite spots and fish until dark. My father had made a hundred copies of my map, and when I'd get home, or the next morning, I would dutifully record what had happened out on the lake: what the weather was, where the fish were located, what lures they pursued, and other miscellaneous "comments." By the time I left for college I had a stack of these maps tucked away in my closet—my fishing logs. The idea was that I could return to them year after year and better understand patterns: when vegetation emerged, how the fish moved around the lake, and so on. But I'm not sure I ever went back to them, after I finished those maps. Instead, I just kept on fishing that lake.

Somehow I managed to save the original map. My father must have preserved it for a number of years and then passed it along to me at some point. I have it folded up in a hardcover copy of my first book, *The Textual Life of Airports*. Here it is:

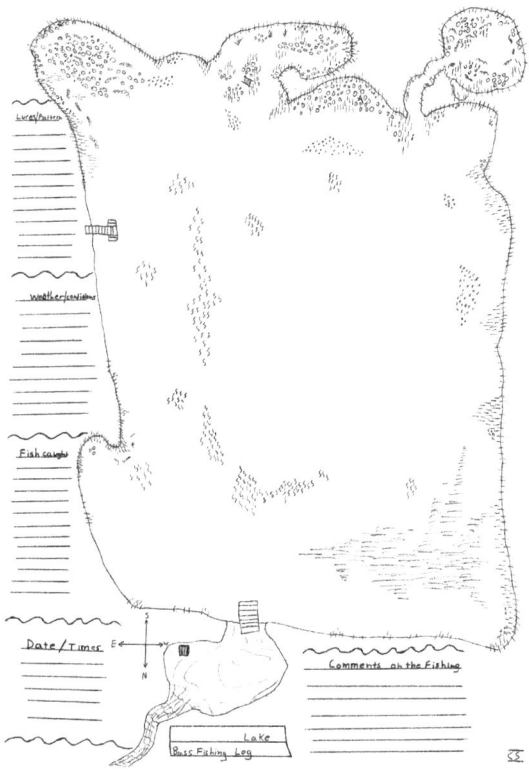

I wasn't even fly-fishing, at that point in my life. I was using a high–gear ratio baitcasting reel and short heavy-action rod, big garish wooden top-water lures I'd order from Oklahoma, and various rubber worms that would sink slowly toward the bottom. (You can see the map is called my "Bass Fishing Log.")

Fly-Fishing

But two decades later, I would return to this lake and adapt these strategies for fly-fishing, and I was delighted to find that my mental map of this lake had remained vividly intact. On some level, the lake meant the same to me as it had twenty years prior, and the fishing had an analogous purpose and focusing capacity in my life. In other ways, I was an entirely different person by that point, and I understood fishing differently, too—not to mention the world and the chaos of contemporary life.

I like fishing alone, most of all. But fishing friends have passed in and out of my life at various points.

My earliest fishing friend was an older guy—in his early sixties?—named Leroy. Leroy's wife was the nanny or something for some children who lived near my house in the neighborhood with the ponds. One day he must have been picking up his wife after work and saw me fishing—or maybe *he* was fishing off one of the docks, and I saw him—and we started talking fish, lures, and techniques. I don't remember if it was after that first meeting or the next time, but he asked me if I'd like to go fishing with him on Lake Lansing. I jumped at the chance—at that point my fishing consisted entirely of the two nearby ponds, and I was eager to expand my horizons.

Looking back, it's baffling to me that my parents simply let me go off with this total stranger at 5:00 one Saturday morning. Leroy pulled up in his massive old Buick, which was pulling a trailer on which sat an old fiberglass motorboat. We drove to Lake Lansing, probably thirty minutes away from my house, and Leroy backed the boat down the launch while I held the bowline. We put our rods and tackle boxes into the boat, and Leroy placed a Styrofoam cooler in as well.

Turning on the ancient Mariner outboard, Leroy aimed the boat away from the launch and cut the motor after just a few minutes. We set up two of our rods with K&E Bass Stoppers: these are rubber worms that come prerigged with three hooks running through the body of the worm, ending on a six-inch leader with a loop tied to the end, to which one can attach a barrel swivel. The worms are packaged in little plastic sleeves, and folded in half; this creates a bend halfway down the worm's body, and this causes them to wobble and twirl when retrieved slowly, thus the need for the swivel (so as not to twist the line and lose casting ability and strength). We cast the worms out toward each side of the boat, at angles, and then put our rods in rod holders, letting the worms slowly sink down. Leroy restarted the outboard and proceeded to troll along the shoreline at an incredibly gentle pace. I could see into the clear water, where fecund weed banks were blooming upward from the bottom of the lake; we were trolling in probably six to eight feet of water.

Meanwhile, we would take our other rods and cast toward the docks that lined the lake. Bass often take advantage of the shade and structure of docks to ambush prey, and so this type of fishing was thrilling for the point-and-shoot aspect of it. You could envision a bass lurking by a piling, and if you managed to cast just past the piling and then bring your lure past the obstacle elegantly, there was a good chance that an explosive strike would occur. We used "buzz baits" for this: these are ludicrously ugly contraptions that are constructed on a heavy Y-shaped wire that sits vertically on the surface when retrieved rapidly; here I'll draw one from memory:

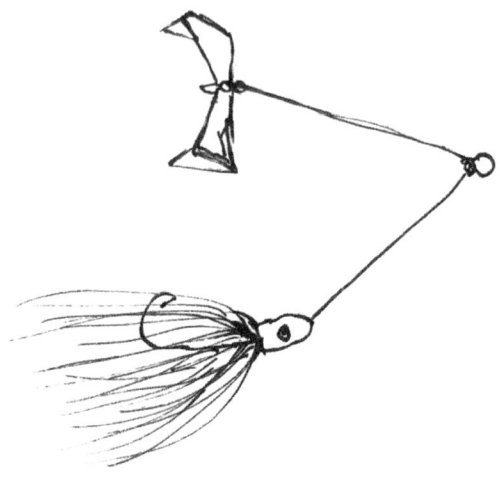

The little "o" on the right is the eyelet that you tie your line to. The lower arm of the V is where the hook is (with a weighted head to make it hang down), and it is surrounded by a "skirt" made of many thin bands of rubber, often with sparkles or two-tone contrasting colors. On the upper arm of the V sits a two-sided blade that can spin on the axis, and it has bent winglets that cause the lure to rise to the surface and chop the water when retrieved quickly. Who knows what this is supposed to resemble: a minnow in distress, disturbing the surface of the lake . . . maybe a duckling, or a mouse or frog racing across the water. Whatever they may or may not imitate, these obnoxiously loud lures would produce aggressive attacks from bass hiding by the dock pilings. (Later in my life, I'd unconsciously adapt these sorts of details to tying flies that skittered across the surface of different lakes.) As we cast toward the docks, every so often Leroy and I would also see our other trolling rods bent suddenly, a fish having taken one of the spiraling worms.

Halfway through the morning Leroy would open up the cooler and pull out two bottles of Faygo pop: rock and rye, I remember distinctly. This was a treat. My parents didn't get us soda, and so the first time Leroy handed me a massive twenty-ounce bottle, my eyes must have bugged out. He always seemed to enjoy giving me a bottle of the syrupy-sweet soda.

I went on these early morning fishing trips with Leroy a handful of times—four, five, six?—and each time we caught many fish, and I was soaking up the exotic (if urban) submarine landscape. The fish were new: we were catching largemouth bass in Lake Lansing—similar to but different from my smallmouths in the ponds.

We'd usually wrap it up and head home before lunchtime. We released all the fish we caught, just there for the experience of catching them. I was always excited to try lures that I had sent away for, spending my odd-job money on fancy new patterns and "fish guaranteed" shapes. But it was always the K&E Bass Stoppers, and the raucous buzz baits, that seemed to catch the most fish. I still have a bunch of Bass Stoppers, and when my son, Julien (then eleven), and I were out in our canoe last summer, I was delighted to watch him catch a bunch of bass and a nice pike on a grape-colored one. He would just wing it out toward the edge of the reeds, and for some fluke reason that cloudy morning the fish just couldn't get enough of it. We repeated the trip this summer, and Julien sized down to a two-hook worm with a little spinner and beads in front of the worm. This one attracted voracious strikes from pumpkinseed sunfish, several of which we took home and filleted and battered in flour and salt and pepper before pan frying them, to Julien's and his seven-year-old sister Camille's delight.

Reading Guy de la Valdène's book *On the Water*, I kept getting annoyed at the repetition throughout the book. *He's already told me that*, I'd think. Or, *Wait, didn't I already run across that particular factoid?* But as I've been writing my own fishing book, I have come to realize that repetition plays an integral part in fishing reflections—not to mention the actual experience of fly-fishing. My fishing is defined by my repeated paths; I literally, if unconsciously, follow my faint footsteps on the lake bottoms and am often surprised to look down and see that I am wading where I had waded days or weeks before. Then there is the tactical repetition that comes with attempting to reproduce an effective fly pattern, or to replicate a cast that landed a fly in just the right spot, a few inches from a giant reed.

Giant reed: that's the common name for *Phragmites australis*, a native but aggressive plant that I love to target when I fish. I learned the name after downloading a guide to common aquatic plants of Michigan, because there are so many slightly different types of vegetation that waver and stand in the water next to me as I fish, and I wanted to know more about these actants. They are so stunningly beautiful, often as gorgeous and mesmerizing to me as the fish I am after, and they coshape the waters together. And they disseminate in repetitive patterns: one type of reed has a catapult mechanism such that when the seedpod dries to a certain bent arc, the husk cracks, the pod springs back upright, and its tiny seeds are flung out in a perfect line. Over the next years, disconcertingly straight vectors of reeds will grow up, marking the trajectory of the seeds. And these reed lines make ideal shelter walls for sunfish, and equally suitable hiding spots for hungry bass and pike.

This morning I didn't go fishing because I thought it would be windy, but instead I woke up to a misty calm—it would have been textbook out there! In lieu of fishing—because we are weaning our toddler daughter Vera, and so I need to be here this morning to take her from her crib when she wakes up—I made a pot of black tea, and I am trying to channel the excitement of fishing into this repetitive prose. No one is awake yet, and I could be fishing, but I am sitting here wading into my mind, casting words onto this white screen before me instead.

I have to make a confession here. Friends and family members sometimes get me fish-themed things as gifts, because they know I love to fish. But I don't really like the fish things as much as I might be expected to. I am not drawn to the fish as a theme, as an image or ornament, as a symbol metonymically related to an *activity* I like—the fish as stand-in for a whole dynamic. These baubles fail to excite me and even cause in me a kind of sad longing, a sharp awareness of the absence of the actual practice. The beer-bottle opener magnetically attached to the refrigerator, with jumping trout all over it. The belt buckle for Christmas when I was twelve, another trout lurching out of a rippling surface. The rainbow trout cufflinks that are too big and garish for me to ever wear. Just the other day, I walked into a little store where my brother-in-law's brother works (it's called the Totem Shop), and after saying hello he smiled widely and pointed out something my sister had said I'd like, the day before: a button-up short-sleeved shirt covered ornately with all types of fish overlaid across one another—and matching shorts.

Even this very morning, after I wrote that paragraph, Camille brought me a homemade Father's Day card decorated with all the different fish I love: pumpkinseed sunfish, bluegill, bass. Adorable. And yet.

I like real fish. Or rather, I like to *fish* for real fish. I like the pursuit, feeling their pull on the end of my line, being stymied just as often, seeing fish elude me, sometimes *not* seeing any fish at all. Fishing is, for me, a kind of deep-down urge—even an ontological necessity. When my niece Ellen asked why I was writing about fishing, I said I wasn't really sure, but it had something to do with fishing being when I feel most alive—and I wanted to think about that, and process it in writing. Ellen looked at me, puzzled. But I know that Ellen, too, has felt the urge: she hooked, then lost, a giant pike once, when ice fishing with a neighbor friend, and still tells the story with suspense and awe.

However, a problem lurks in these sentences and across the pages. The nagging paradox is that it's always about *not* fishing. I'm not writing *as* I fish. That would be the purest fishing book: embedded journalism, or immersed journalism—fishing and writing at the same time. I've even thought I should try this for a chapter: Fish, write. Fish, write. Fish, write. They can't be done simultaneously, but what if they were brought closer together? If, as Jenny Odell puts it in her book *How to Do Nothing*, "curiosity is what gets me so involved in something that I forget myself" (104), then fishing writing might count twice over.

Sometimes I'm relaying a memory from the past, and then I'm in a scene in the present. I can't keep track, from line to line, of where I am in time—in my head or recently on the water. It's a tension that never resolves. Fly-fishing itself relies on tension— some people who fly-fish bid farewell with the phrase "Tight lines." The fly line should be this way: tight, not looped or bowed or tangled. But if the tense of my narrative is a problem, it's a tension that I'm deliberately exacerbating.

A few times each summer I manage to get all the dishes done, the children in bed, the house tidied up—and there are still thirty minutes left of sun before dusk. The air is still, trees unmoving. I quickly and quietly pack up a rod and my waders and take a single fly box and my pliers on a simple loop of old fly line that I can drape around my neck, and I drive to the lake. I'm fishing ten minutes later.

Fishing at dusk on mirror-calm water has a certain magic. The surface dimples with insects landing and fish feeding. Sometimes fish explode out of the water, defying gravity to snatch a dragonfly or mayfly hovering feet above the surface. As the shadows get longer, the water gets more active. Mosquitoes coordinate to form maelstroms around any heat source: namely, you and your head. The fish are hungry, and aggressive.

When I last set out to fish at dusk, on a Saturday evening, I was reluctant at first because of the near certainty of weekend fishers slamming around in boats, disturbing my idyll. I really prefer to bask in all the nonhuman noises of the lake, so as to let my own humanity go. Saturday night in June, perfect weather . . . it was bound to be a carnival out there. Therefore, I was almost shocked to find, when I reached the terminus of the two-track that runs into the lake, that no one else was there. I had the lake "to myself"—a provisional fiction but one that matters to me.

I waded into the lake and started casting toward submerged redwoods and toward stands of reeds, even way out into open water—sometimes toward night, when the water is like this, big fish are just cruising and looking for an easy meal.

A pair of swans were directly across the lake from me, and four additional swans another couple hundred yards down the shore from them. I had never seen so many swans on this particular lake. Eventually the four took off, their giant wings beating in synchronization, creating a turbulent cacophony. Fifteen minutes later, the other pair also flew away. I heard loons crooning at the other end of the lake, but I couldn't see them.

I had made the mistake of bringing only my stoutest eight-weight rod and a box of very large poppers and other surface flies: in my mind I had seen this as a perfect night to get some gigantic bass from out in open water, which is easier to cast to and detect strikes in as darkness falls. You just wing it as far out as possible and slowly chug the popper back in. It looked great for this, but I wasn't generating any interest.

Having no luck out in the open, I turned and targeted a bank of sedges near the shoreline—and right when my fly landed, it was engulfed, the water erupting in a huge swirl. I set the hook but too late—the fish wasn't on. I immediately swung the fly back to within a foot of where the strike had occurred—they will often hit it again, even a third time, if they miss—but nothing. Even if I could have gotten the fish to strike again, my cast was so long that there was too much line out. Plus, halfway between me and the spot I was aiming for, a dense stand of reeds had my line draped sloppily across them, like a raggedy clothesline—not ideal for setting the hook, much less playing a fish. It was not pretty, hardly precision fishing.

In any case, this glimmer of action caused me to change my tactic, and I started to cast toward shore, finding that the fish were in the shallows after all—and that they would explode out of the thick vegetation toward my fly. This got my heart thumping, but I also realized that many of these fish were toothy pike, and (a) I had not brought any steel shock-tips (strong metal line for the last foot or so, to guard against pike's teeth, which will slice through the toughest monofilament), and (b) it was rapidly getting darker, and removing a hook from a snarling pike in inky water in low-light conditions is dicey. The thought of it terrified me, in fact. Remember, I had intended to catch bass. Still, I couldn't resist the danger zone and kept casting toward the cuts in the weeds and downed trees where I suspected the fish were lying in wait. (Bass? Probably pike.)

Around this time I heard a low rumbling coming across the meadow, and the telltale sound of branches from an overhanging tree scraping across a rooftop. A tall and narrow white van lurched into view, one of those Euro models that have become popular more recently in the States. Ugh, people. They parked, and I heard dogs jangling their collars as they dashed around. It was twilight at this point, and so I couldn't make out too much, but it looked like a thirty-something couple—#VanLifers, maybe. A big husky-like dog ran down the shoreline toward me and started barking, and a heavily bearded guy strolled down and said, "Aria, come! . . . Sorry, bro!" I said something innocuous like "No worries!" And then he stood there on the bank and watched me cast.

At one point he said, "Whoa . . . that's a hell of a cast!" I was slinging my flies way out, farther than was really effective, to be honest, but with such still water I needed to get the fly out beyond my own disturbances. Maybe my cast looked impressive from the shore, but I'm *not* a graceful caster. I've had no proper training and have just improvised since I started fly-fishing. There are all sorts of supposed rules as to the proper way to cast a fly, but I don't follow them (or even remember them, if I've learned them at all). It was kind of the guy to compliment my cast, but I felt sheepish, since I *know* my casting isn't elegant. Still, it gave me a little confidence boost.

Right then I finally hooked a fish that came slamming out of the weeds, and as I fought it back toward me, I heard the guy shout, "Fish on?!?" which I found funny because it was very obvious to me that this was indeed what was happening: I had a fish on. And the guy was now watching live TV after getting out of his van. I had brought the fish almost close enough to grab when it charged into a dense cluster of reeds and threw the hook; I saw its gaping mouth for a moment before it got off. It hit like a pike—like a torpedo—but it fought like a bass, and its mouth looked bassy when it finally shook free. I didn't make a big deal out of losing the fish, since I had an audience. For all he knew, I caught it and quick-released it. The whole thing was a debacle from where I was standing in the dim light—but from the guy's vantage point it must have looked like a high-definition fishing show. He bellowed, "SO COOL!"

I spent the next endarkening hour trying to catch a fish—and I kept losing fish. I couldn't quite set the hook; once a fish even completely ripped off the head of my hand-tied popper, leaving just its tattered tail (I found the shredded head floating a few minutes later and retrieved it; after I finish writing this morning, I will repair it). As things got more interesting, it also got murkier and scarier out in the water. It becomes easy to imagine impossibly giant creatures swimming around my legs, or slithering water snakes of epic proportions approaching with appetites.

The thing about fishing after dark is that everything gets more difficult. It's harder to gauge distance, and casting becomes a kind of leap of faith every time—that you didn't hang up in weeds or on a log (I did and had to go unhook my popper by hand, blowing an otherwise really promising spot), that you might have managed to land your fly close to a piece of structure, that you'll be able to see a strike, or feel it . . . And the bottom of the lake becomes a horrorscape, full of spontaneous chasms and unsettling textures. What was familiar becomes radically defamiliarized. It's a different place.

I didn't touch a single fish the whole evening. Meanwhile, the #VanLifers had retreated to their mobile enclave, dogs and all. At first I thought they were having a fire on the shore: I saw a smoldering glow in the sand as I waded back toward my car. But as I got closer, I discovered that it wasn't the embers of a fire: it was a red blinking LED light on some sort of device. Next to it was a small antenna-looking thing with a soft blue light. What the hell? I had no idea. Maybe a Bluetooth satellite antenna? I glanced over at the van, and the blinds were all down, but light was dancing behind the curtains. Watching a movie or show, perhaps. I tramped back to my car, got out of my waders, and drove home in my own dimly illuminated capsule.

In her book *Unthinking Mastery*, Julietta Singh describes "a *dehumanist* necessity" (15) in the titular problem: to unthink "mastery" as a concept, one has to be willing to let go of some of the primacy of the human. Other things, other entities, take on greater importance here. Singh's scholarship "reaches toward other modes of relational being that may not yet be recognizable" (16). This is what I find when I fish at dusk: a dehumanist necessity. A tuning into things beyond me—and not to master them but to merely coexist here.

On the one hand, this is a body of water that I know quite well, and I understand what fish species exist here and how to occasionally tempt them into biting my fly. I have a decent cognitive map of the bathymetry of the bottom. I am confident following the shoreline and know how to get back. But on the other hand, I quickly become not the center of things. I am minimized. Other creatures and eccentric zephyrs take precedence. I am not the master of this place.

I didn't catch any fish that night, but I captured other material: the stuff that became this piece of writing. Perhaps this establishes a method. When fly-fishing, I'm not always achieving the proper goal but finding something else along the way. Of course, I could never admit this to myself when preparing to go back out on the water.

There's an old adage that goes like this: If you give a person a fish, you feed them for a day. If you teach that person to fish, you feed them for a lifetime. This is generally a sound tidbit of wisdom: an aphoristic paean that enfolds self-sufficiency with mentorship and community. And it is low-key outdoorsy.

But what if you fish not to eat but for some other ineffable reason? What if you're a bad teacher?

Today I took my son, Julien, fishing. We were going to try him in waders for the first time. I've taken him out on the canoe plenty of times, and we've fished from docks and shorelines . . . but waders are a different experience. I wasn't even going to teach him how to cast a fly rod but just equipped him with his old Zebco spincaster and trusty K&E Bass Stoppers. The point was to get the feel of being *in* the water, walking in and through its cryptic underrealm.

It was a day just after a cold snap, a brief few hours of calm before another microsystem blew in across Lake Michigan— we watched it approach, lightning fracturing the sky, as we hurried back to the car later.

But the fishing itself was slow and unproductive. Julien was frustrated that he didn't catch any fish, and I was frustrated at myself for being such a terrible teacher. I was trying to teach him how to walk a few paces before casting, then to stage himself and remain still after casting, while focusing all his attention on his rod and line and lure. Teaching him a slow retrieve, the gentle twitch of the artificial worm, visualizing the making-alive of the bait: aiming for a reed stand or a dark drop-off, then bringing the worm back in slowly, creepily. It's where you cast, then how you draw it temptingly back to you.

Fly-Fishing

I was teaching him to cover the water in front of him, starting close to shore at the edge of the vegetation and then gradually fanning out toward the deeper reed lines. Teaching him how to lead a fish: how to cast several yards in front of fish so as to not spook them. And how to let the worm drop after it hits the water, descending slowly as the ripples settle above. It all sounds so good, in the abstract.

The reality was me shouting, saying things sarcastically to him, offering snide feedback on every cast, critiquing his posture and his grip—in short, being the worst teacher I could imagine. As I reflected on it later in the afternoon, I thought: How could I teach *anything* effectively, being so impatient and generally awful?

Julien presented me later with a drawing he made, which had the banner "Go fishing" and as a subtitle "It's fun." The picture depicted a wild-eyed fish holding a fishing rod and casting a worm, and a human in waders and a fishing shirt swimming underwater—vulnerable and dumb looking. Two smaller fishmen swam nearby, wearing miniature waders. Reeds stood out of the water, and on one was a dragonfly newly emerged from its larval exoskeleton; we saw those in the morning, and he'd represented it with surprising accuracy.

I didn't teach what I set out to teach. My lessons in fishing had gone horribly, and Julien was understandably annoyed at me while we fished. But he had no doubt learned a plethora of things that I wasn't even tuned in to. In his mind he inverted the normative image of fishing: *we* were the susceptible prey.

An alternative title for this book could be *The Tangential Poetics of Fly-Fishing*.

Terrible fishing day yesterday. When I awoke at ten minutes before 5:00 a.m., I heard wind in the trees—it had rained all the day prior and into the night, but it looked like there *might* be a lull between storms. I convinced myself of this remote possibility. I chanced it and headed out to one of my favorite lakes. I had packed only a lighter-weight rod and a streamlined box of flies—I had a specific shoreline in mind to fish and exactly two ways I was going to fish it (one fly pattern for the surface, another for the subsurface).

When I arrived, though, the wind had shifted—and increased. As I approached what used to be an old boat put-in, a foot-tall wall of foam barred my path. It extended out into the shallows for about ten yards, having been blown in by the all-night wind. This phenomenon occurs sometimes in these silty shallow bays, when all the bacteria get whipped up and frothed by the waves.

It was still dark, and I experienced a brief spell of epistemological vertigo as I stepped into the foam—what was this substance I was moving through, neither liquid nor solid nor gas? How could I know?

I should have accepted this as an omen, or a sign from some ethereal deities of the dunescape. But instead, not believing in those gods or any others, I waded out into the grayish predawn light and began to do battle with the gusting wind. The waves rocked me in my waders—making me uncomfortably aware of how calm it usually is when I fish. Now the water was splashing up toward my chest, and the boiling crests made my initial top-water fly laughably invisible to me, and even more so to the fish. It was futile. I spent that first critical hour roving along the shoreline trying to find a protected nook in which to fish, but the wind seemed to follow me everywhere. Frustrated and already exhausted from casting against quick-reversing microgales, I decided to hike back to my car and head to another nearby lake whose southern cove I knew would be protected at this time of day.

Wading into the new water, it looked right: the wind overshot the tree line, and the surface curving before me was still. The overcast sky created the perfect lighting. I should have been here an hour ago! Still, I had a few hours available to me, so I made my way to the spots where I knew fish should be. This was the same place of my dusk fishing misadventure a few nights prior, and of my botched instructional outing with Julien more recently. I return again and again to these lakes, as if proving Jenny Odell's observation that "nothing is so simultaneously familiar and alien as that which has been present all along" (125). Same place, everything different.

But the disaster of the first lake followed me around, and I missed fish, or spooked them as I approached their redoubts. It was a morning of setbacks and missed opportunities, and I drove home later feeling as though I'd squandered everything.

Scrolling through the photo album on my phone, I can see the hundreds of pictures I've taken of fish and fishing spots over the past several years. Visual slices of time. Some fish are held in my palm or stretched across my arm or a thigh. Landscape shots show hillsides, reeds arcing out of the water, mist tumbling across tree-lined ridges, and sunbeams dancing across ripples. Sometimes I try to hold the phone out as far as I can and take a selfie as I hold my catch, but these shots look forced and do no justice to the fish or anything—except maybe the phone itself.

On my first fishing outing of the season this year, I caught a large pike and just had to snap a photo of it. This pike was an armful, and it still had the fly in its mouth: a long streamer I tied, olive on top and white beneath with red 3D eyes. Instead of taking the hook out first, I decided I needed a picture of this pike immediately; it was formidable. I unzipped the chest pocket of my rain jacket, snapped a few shots of the pike—it was hard to fit in the frame, long like an em-dash clause out of control, held in my right hand, its broad tail fanning out against my knee— and slipped my phone back in my pocket before turning back to the hook in the pike's mouth. As I reached carefully toward its jaw and teeth—rows on rows of razor-sharp teeth—I felt the slightest lightening . . . like an imperceptible burden had been released from me.

Then I saw it: a silver rectangle feathering through the green water, toward the silty bottom. My phone! I'd forgotten to zip my pocket back up! I knifed my hand into the water and managed against all odds to snatch the device before it settled into the unfast substrate below, where it would have disappeared for good. I gave it a shake and shoved it back in my pocket, zipping it shut this time. I removed the fly and released the bemused pike, who jetted off into the depths. The whole incident probably took less than ten seconds, but it replays in slow motion in my recollection.

Later I put the phone in a Ziploc bag full of basmati rice and left it overnight. It was fine, and I have just finished thumbing this story into the very phone that swam for a few moments in the water of the pike, as I held the pike in the air that is the natural habitat of the phone.

All fish photographs bear an unusual trace: this wrongness, an out-of-placeness. A wrongness that is nevertheless tempting, to capture the fish out of its element. Even the best fish photographs tarry in this uncanny realm. Perhaps this is why fish never look as big or as beautiful as they really were, in the water, when we cradle them briefly, before the return to a place that exists beyond any frame.

Often, the night before fishing, I can't decide which lake to go to. It all depends on wind conditions, precipitation, and my last fishing spot. I like to vary where I go, but at the same time, some shorelines beckon for multiple days in a row, when I suspect that fish are congregating or amassing for a period of time. The indecision vexes me and can continue to vex me right to the point when I've committed to a place . . . when I've passed one road or turned down another. Chosen one shoreline to wade or headed in the other direction.

I've fly-fished in Montana, which I'll talk about later, and I also spent a few years fly-fishing on the Mississippi River at the bend in New Orleans colloquially known as "The Fly." This was when my family and I lived in Uptown, near my job at Loyola University. I could walk to the river and fish the sunrise and then be home to make breakfast. That fishing wasn't particularly productive, but it was always full of mystery and surprises. I even caught a few fish throughout the years: a school of needlefish once appeared in an eddy and went berserk for an old Mickey Finn, and a catfish improbably snatched a Prince Nymph bouncing along a sandy bottom. Here's how it happened:

When I moved to New Orleans, I didn't plan on fishing. I was beginning a tenure-track professor position, and I spent the first couple years focused solely on teaching my courses and revising my doctoral dissertation into a book. And my partner, Lara, and I had our first child—those were very busy years.

But one morning when I was taking my then-infant son, Julien, for our usual three-mile walk along a big curve of the Mississippi River near our home, I noticed a guy standing up to his knees in moving water that resembled a Wendy's Frosty, casting a fly rod. He wasn't wearing waders; he had no fancy fishing vest on. Long dreadlocks hung down his back, and he wore a ragged Grateful Dead T-shirt. His madras cargo shorts were tattered, and I could make out the shape of a fly box in one of the thigh pockets. I was intrigued. After that, I'd look for him each day, and I gradually pieced together his route: he would work a sandy stretch of the river in the morning hours of certain days.

Finally, one day I worked up the nerve to scramble down the riprap and talk to this weird urban fly fisherman. Julien was along for the ride, tucked into a baby carrier strapped on my chest. I asked the guy if he was having any luck, and Brian—that was his name—went into a detailed explanation of the variety of fish he'd caught in the river, how he was fishing for them, how he had a 10 percent success rate, and how he planned to try some new fly patterns when the speckled trout would appear. To paraphrase an expression from Ernest Hemingway, I began to feel all the old feeling.

I went home that day, got out my gear from the attic, and rigged up my rod. I sorted through my tattered Montana flies to see which ones I might repurpose for this new mystery water. And the next dawn I was out there with Brian, fly-fishing in the Mississippi River. That first morning I caught a ladyfish and a white bass, and I was ecstatic. Brian and I became occasional fishing buddies, meeting up sporadically whenever our schedules overlapped and trying out different flies and techniques, wading this ominous river bend with impressively low yields but always worth it for the surprise catches and other eerie riparian discoveries. Alligators in the shallows, grass shrimp skittering over our toes, the stain of the river on the skin that takes several days to wash off. This is what the philosopher of science Donna Haraway might call an adventure in the "Chthulucene": a tangled-up encounter with sludge life that becomes impossible to isolate, for it runs everywhere and gets into everything the closer you look.

As our family grew, and as real estate prices in Uptown escalated beyond the reach of my faculty salary, we decided to sell our small single-shotgun home and move to Mid-City, where we could get twice the house at half the price. But then my fly-fishing adventures on the Mississippi abruptly came to a halt. I saw Brian only a few times after that, when we aimlessly fished the vast shoreline of Lake Pontchartrain and caught nothing. He later texted me a photo of a massive drum he caught there; he persevered, and figured it out. I hope to run into Brian again one of these days and tell him what I've learned about fly-fishing around our new home, in the Old Basin Canal and in Bayou St. John, catching Rio Grande cichlids and redear sunfish, especially—even if Brian pokes fun at me for my interest in catching the smaller fish in our region.

There is an irreducible sadness to fly-fishing. Fly-fishing is decidedly effortful, a constant tussle with gravity, air, and water. Sometimes, sure, it looks and feels graceful. But there is a tension always present, an interior pull that drags one down. Something about the contradiction between the rambling mind and metronomic casting. Even the best fishing days are lined with a deep melancholy. Even if it only becomes apparent at the end of a trip, on the return to everyday life back on land.

I have been resisting writing about one particular morning because it is rife with details, and I was sure I wrote about it already in an email to my friend Greg Keeler, whom I tend to keep updated with respect to my fishing adventures. But for the life of me I can't find any trace of this story, even after scouring my computer hard drive and Dropbox and email inbox for key words: "gar," "alligator," "Jeep," "Hula Popper," "dead mermaids."

    Here's what happened. It was September at the beginning of the school semester back in New Orleans, and I was missing fishing so much that I drove two miles to City Park before dawn to cast into the lagoon. I had found this one area near a playground that looked good. While my children played, I surveyed the lagoon: water hyacinths clumped everywhere, and yardstick-long gar cruised right under the surface. Turtles were all over downed trees stretching out into the water. Plastic water bottles and energy-drink cans bobbed among the aquatic vegetation. And snail shells the size of baseballs were scattered along the muddy bank. It looked fishy.

When I walked to this spot in the dark, though, everything was different. Each sound made me jump. I started casting a big fly out into the open water, because I couldn't make out the structure near shore, and I didn't want to get hung up. As I cast into the first faint glow of morning light, I noticed an odd reflection on the water across the lagoon: it looked like people. People hanging. On gallows. One was hanging upside down—hanging by a . . . tail? As the oily sun came up through the live oaks, I started to make sense of the grisly scene. It was part of a "scream park" set up for Halloween—an interactive spectacle of terror and gore, with various dioramas set up to conjure different atrocities. In this case, it was slaughtered life-size mermaids hanging from crossbeams. It was nauseating, seeing it out of context and having to figure it out without a trigger warning.

I caught a nice, very dark bluegill on the surface, which made me happy, but I couldn't shake the creepy feeling of having been startled by the scream park. New Orleans is too loaded of a place for artificial terrors. It was just after seven o'clock. The gar were moving. I had tied some long, sinuous flies for gar, but I really didn't know the technique for how to get them interested. I would cast past them and swim my flies next to them, but the gar would simply disappear gracefully into the depths. I love gar, our nearby prehistoric cousins.

At that point a cherry-red Jeep roared up and screeched to a stop in the grass behind me. A teenager jumped out, grabbed a bright caution-green baitcasting rod and matching reel from the open back, and marched down to the water, standing about fifteen feet away. He cast what looked like a giant Hula Popper way out toward the opposite bank and started chugging it in with vigor.

I said, "Fishing for bass?"

"Yes sir! I'm in a tournament right now. I already caught three: two one-pounders and a two-pounder."

"Nice! Did you catch them on that popper?"

"Yes sir!"

He cast a few more times, fanning across the expanse right before him, pulling the popper in rapidly with resounding blurps and bloops. No fish here. Just as abruptly as he showed up, he reeled in quickly and charged back to his Jeep, threw his rod in the back, and tore off. The whole encounter lasted less than five minutes. Kids these days!

I thought, *Well, if he caught bass on that thing, I've got to size up.* I was using my six-weight, not great for casting big poppers, but I tied on a size 4 black BoogleBug and cast parallel to shore. I know where bass hang out. I had been casting blind, using smaller surface flies and bigger streamers, because I didn't really know what I was after. But I know how to catch bass. I gave the popper a twitch, and WHUMP. It was gone, and I had hooked a leaping, lunging sixteen-inch bass. I brought it in, snapped an awkward arm's-length-production photo of myself holding it, and released it. I had figured something else out, this morning. Something better than dead mermaids.

I kept fishing. No more action, though I did see lots of ripples on the water and predictable, if always startling, mullets leaping repeatedly every minute or so. Mullets seem like they'd be fun to catch on a fly rod, if you could somehow imitate whatever minuscule detritus they hoover near the surface.

A large log seemed to be drifting toward me. I cast aimlessly, mentally going through the fly box I had brought along, wondering if I should try anything else here, this morning. The bass hour had passed.

I had the sudden sensation that someone was watching me. Then I heard breathing. Loud breathing. Right in front of me. (It also turned out that my feet were being devoured by chiggers; I would discover this later that night, when my toes would puff up into red itchy prostheses.)

It was no log. It was a twelve-foot alligator, and it was mere feet away from me now, staring at me and exhaling forcibly out of its snout. I inched away, reeled in my line, walked nervously back to my car, and drove back home. Later that day I tied some new, even larger black foam poppers. But I never went back out at dawn to use them at City Park. I saved them for Michigan.

When I tell people I am writing a book about fly-fishing, often they will ask if it is going to include pictures. I think they usually mean photographs, but they could mean illustrations or diagrams, too. It's curious that the genre of the fishing book also comes with an expectation of some—even if minimal—visual aids. Mark Kurlansky's book, for instance, has little sketches that are his: scenes of different places that he mentions throughout the book. And simple drawings of flies between the chapters. It also includes all sorts of historical images: ads for old fly rods and the like. But these are not directly discussed, just there to add to the ambience, as if to be ultraclear that THIS IS A FISHING BOOK.

I debated whether to include images—I have plenty of my own from over the years, as I mentioned earlier. And some would make vivid the places and fish I am writing about. For instance, I snapped a blurry shot of the first fish I ever caught on the Mississippi River, a ladyfish—though I had no idea what it was, in the moment. It's flopping on the riprap and wire meshwork that bespeak the effort to hold back the giant river; part of my reel is visible at the edge of the frame, small shells scattered around beneath. The fish was just about six inches long, nothing to brag about. But I was tickled by this catch: fly-fishing had worked, unbelievably, once again. Yet when I look at the photo, something about it dissatisfies me, even as it triggers a positive memory. I share my iPhone-snapped pictures with friends, but they almost make no sense out of context. The truth is that I often feel foolish even as I click "send" and the digital photos zip off into the digital ether.

Photographs feel incorrectly representational, for the story I'm trying to tell. They can risk seeming decorative—and like I'm drifting away from actual fishing. Which, of course, I am. But I want to keep the *drift* toward the writing, not toward pictures of waterscapes, or even fish.

Fishing is often so good when it's raining lightly. But the problem is that when it's raining, my family is often homebound, and I am needed at home to monitor the corollary tempests of children building forts, stacking pillows, playing dress-up.

One afternoon in Michigan, Vera was napping, and my father had taken Julien and Camille into town to get ice cream and go to the playground—there was a break in the rain after four days of near-constant downpours. Lara told me to take advantage of the two hours and go fishing—and within five minutes I was racing toward my lake.

I found fish in the tall reeds sixty feet offshore, and I waded the length of this reed line and cast into the pockets, catching many large bluegill and several football-shaped bass. It drizzled on and off, the light conditions low and perfect. I heard the loons in the distance, and sandhill cranes. Kingfishers fished along with me, closer to shore. Writing about these birds now reminds me of a scene I wanted to include, from when Glen and I were fishing a couple weeks prior: we were wading through the murkiest, muckiest corner of a lake when I looked down at a stand of four reeds and saw a nest intricately placed between the stalks: it was a redwing blackbird's nest, with four eggs in it. I carefully waded away from it.

I haven't explained Glen yet, though I've mentioned him a few times. Glen and I met at the local food co-op, where he works as a produce manager. But he'd spotted me at the fly shop before that, and so one time when I was doing our monthly grocery run, he approached and said, "Did I see you at Northern Angler?" We started talking about fly-fishing—but to be honest, I am pretty guarded about this topic normally, since my style of fly-fishing is not exactly the popular kind. But soon enough we both realized we liked the same kind of fly-fishing: for bass and bluegills, especially. Glen has a son, Rhine, a little younger than Camille, whom I'd remembered seeing at the fly shop with him that time.

We would share fishing stories over the following year or so, whenever I'd be in town doing the shopping and he was there. Eventually we made a plan and met up to fish. I took him to the lakes I knew so well—it was unfamiliar territory to him, as he lives about an hour away, on the other side of town. Over the years we became close friends, fishing once a week during the summers and texting throughout the school year across the country. Glen ties the most perfectly sculpted flies I've ever seen, and they always catch fish. I get tips from Glen about how to tie flies, but mine always look like gutter trash compared to his works of art. Fishing is our main point of connection, but we talk about parenting a lot while we're fishing. And *Star Wars*.

Once we had waded into a slough where I knew the bass would be, and I was sort of narrating to Glen how it was going to go: put your popper *exactly in that one-foot diameter spot next to that downed tree precisely in front of that lily pad, yes, right there, good... The bass is checking it out! Wait, wait...* And the bass engulfed Glen's popper, and after Glen brought the fish in—I was going to say *fought the fish*, but I changed it to *brought the fish in* because another thing Glen and I have in common is that we actually love the fish and consider them no less equal as beings or things, and so we are extremely careful and gentle as we fish, even admitting (and trying to reflect on, and work through) the paradoxes of violence and physical wear involved therein—he said something hesitantly about how I used the Force, and I said, "Oh yes, it's just like that." And we started talking about the new *Star Wars* films, and various characters and plot points, vintage *Star Wars* toys ... and at some point during the conversation, Glen said, "I have to admit something: I'm a big nerd."

I said, "Glen, I'm a fucking *English* professor!" It doesn't get much nerdier than that.

And our friendship then grew to encompass *Star Wars* and Legos and the Marvel Cinematic Universe . . . subjects we texted about throughout the school years as we parented and shared these things with our children. When we fly-fish together, often in the rain because we both love fishing in the rain, we also talk a lot about how to parent in this precarious world, how to recognize our own limitations as parents, how to raise children who are caring and tolerant and attuned to a planet in peril, a planet they might make a little better during their time here.

Having repaired the snapped thwart on our old Grumman canoe (straight-line windstorm; canoe up in the trees), I took Camille out for a quick fish: we had an hour or so while Lara took Julien and Vera into town to pick up groceries and visit a new playground. It was an overcast day, no one else on the lake. Once on the water, Camille was eager to cast the K&E Bass Stoppers toward the reeds, and she quickly caught a small bass and then a small sunfish. She was thrilled. I did not bring my fly rod that day. I was focused on getting Camille to catch some fish.

Having not had parents who fished, I'm extremely leery of ever foisting this practice on my children. I want to give them the opportunity, but I never want to push it on them or make it seem obligatory. I recognize that most people have zero interest in ever fishing, or even if they try it once or twice, it will not grab hold of them the way it did me.

Anyway, half of this trip is just being out on the water together and with the rest of the lake—the fishing was only a part of it. When Camille heard the cry of the loons across the water, her face lit up.

When Lara pulled into the parking lot an hour later, we paddled back to the launch, and Camille jumped out of the canoe eager to tell them all about our brief trip: the fish, the loons, a soaring heron. Time had expanded, in those sixty minutes: so much rich detail, so much teeming life on panoramic display. Meanwhile, I saw little Vera eyeing the canoe, and so I said, "Vera, do you want to go for a ride?" She nodded vigorously. We stuffed her in the smallest life jacket, and I plopped her on the front seat of the canoe, backward so she could hold the thwart. "Now hold on tight!" And she did.

I paddled just a short distance out into the calm water, and two loons appeared close by. We drifted near them. Vera was mesmerized. After a few minutes I paddled back in.

We got all the gear back in the car, and I loaded the canoe onto the roof rack.

Later that night, when I was getting Vera into her pajamas, she looked at me and said, "Boat! Hands, tight!" She pretended to grip an imaginary thwart.

Fly-fishing was a present absence this day, a possibility in reserve, suspended in lieu of other people and experiences.

I'm naming things inconsistently. Some people have last names, and others do not. Part of this is out of respect for privacy: Brian is also a brilliant illustrator but was very protective of his work and worried that someone would use it without his permission or without due credit—of course, Brian's concerns are justified, as a Black man living in the fraught port of New Orleans, under the long shadow of slavery and discrimination. Leroy is also given only a first name, but that's mostly because I only ever *knew* his first name. Greg Keeler gets a last name, because he is an author. As do the various other writers and thinkers I cite. Ted doesn't get a last name, because I didn't tell him I'm writing about him—and I don't even know how his story ends, at this point. The carp will get its Latin name, and so does the giant reed—maybe because I wanted to bestow on these a certain extra intrigue—but other things just their common names: bass, pike, bluegills, lily pads, white pines, pondweed. Glen is just Glen—in the best sense of a simple appellation. I do not name the lakes I fish in northern Michigan, even though they would be easy enough to recognize and find for anyone in the area. But I name Bayou St. John, in New Orleans.

At one point in *Why Fish Don't Exist*, Lulu Miller suggests that even the name *fish*, "in a certain sense, is a derogatory term. A word we use to keep ourselves comfortable, to feel further away from them than we actually are" (181). In other words, while taxonomic names suggest mastery and knowledge, the broader category of fish in fact weirdly distances humans from their watery kin. I thought of this as Julien and I observed a pregnant three-spined stickleback, *Gasterosteus aculeatus*, in the stream near our home, and as we learned about its reproductive methods and patterns. And how it was related, circuitously, to pipefish and seahorses! The more we studied the one-inch-long but very plump stickleback, the more familiar it became to us—we didn't even see it as a "fish." Naming this smallest fish opened up a network of delightful relations, a network that caused the name and even its otherness to recede.

I was reading Octavia Butler's story collection *Bloodchild* during the summer that I was working on this book. As I started the title story, I immediately tripped over its opening lines: "My last night of childhood began with a visit home. T'Gatoi's sister had given us two sterile eggs" (3). The first sentence is clear and distinct; we know where this story is going. But then the unusual proper name in the second sentence, followed by the sterile eggs, completely confuses things. Butler uses the name T'Gatoi continually to destabilize the reader's sense of what is happening, even as we become more familiar with the character over the course of the story. It's an unname and an ur-name, a practice of naming that draws attention to how strange names in fact are. It confused me in the moment but helped me think about how we name, when we name, and what lies beyond all names.

Another alternative title for this book could be *Fly-Fishing in the Same Lakes More Than Twice*.

Today I took Ted out to scout the place where he wants to fish on Monday—a lake that used to be easier to get to, until the National Park Service closed off one of the access points for ecological restoration, and high water submerged the only other road that led to the water and made it impassable. But I knew about another trail, and I convinced Ted we should try to put in there. He moves slowly these days and wasn't sure he could even get to the water. He also wanted to use the wooden boat that he had designed and built himself: a nine-foot flat-bottomed craft that resembles a cross between a miniature river dory and a johnboat. I told him I could spare an hour to scope out our approach.

Ted picked me up in his pickup truck right on time, and we drove to the lake. At ninety-six, Ted can still drive—but I found myself holding my breath around the curves.

We got to the spot, and I showed him where to park, and we made our way step by step down the partially hidden trail toward the lake. I fashioned some simple walking sticks for him partway down the trail, to help him along. What normally takes me about two minutes to walk, from car to lake, took close to twenty minutes.

At the old put-in, Ted realized that he could do it—he really hadn't been sure he could walk this distance. So we *could* fish here—*if* we could get the boat down to the lake. I was ready to head back home—I had a narrow window—but Ted scratched his beard and said, "You know, Chris, I'd hate for us to get here on Monday morning and have a snafu. I think what we should do now is make sure we can get the boat down here."

I nodded. "All right, let's do it."

We traced our steps back to the truck—another twenty minutes—at which point Ted guided me through how to un-strap his byzantine lattice of ties and mechanisms that kept the little boat secure in the bed of his truck, and how to carefully set it down on a set of canoe-towing wheels that he had retro-fitted to accommodate his boat. The boat is a work of art, and the hodgepodge of bungee cord rigging made for a laughable juxtaposition. We turned around and headed back toward the lake. The boat was light and easy to maneuver, and I got down the trail with no problem. Ted followed behind, moving at a pace slightly faster than a snail.

I was ready to turn around again and haul the boat back to the truck, but Ted was scratching his beard as he pondered the glassy surface of the lake.

"Chris, I think what we ought to do now is make sure you can handle the boat, see how it feels on the water."

I checked my phone for the time. Lara was going to be *pissed*. "Okay, Ted."

I eased the boat off the wheels and glided it into the narrow channel. I climbed in. Now, I'll admit that I *had* been wanting to row this boat ever since Ted told me about it, several years before. But I was not prepared to do this today. I had no life jacket, and I was not really wearing what I'd normally wear to take a boat out. But I shoved myself out of the cedar-skein channel, and then, pulling myself past the last overhanging limbs, I put the oars in their locks and rowed out to the edge of the reeds. I saw a fish rise twenty feet away.

I rowed back in, assuring Ted that the boat handled beautifully and that we'd be all ready for Monday. He was beaming but clawing at his white beard again.

"Say, Chris, now let me climb in and you can see how it will be *easier* to row with two people in the boat . . . "

It took Ted a good minute to crawl into the boat and get himself seated. And then I had to step out of the boat and coax it back into the deeper water. I was wearing jeans and sneakers— Lara and I had planned to go to a farmers market when I got back, and I was dressed for that. Not anymore. I climbed back in and sculled the oar off the stern, pulling us back toward the open water.

Did I mention that Ted had his Orvis Helios rod all rigged up and laid across the thwarts of the boat?

Once we cleared the reeds, Ted said, "Now, Chris, pick up that rod and catch a fish."

Lara was going to kill me.

I unhooked the chartreuse popper from the eyelet, cast toward the reed line, and handed the rod to Ted. It was an old popper and wasn't floating, so I had Ted bring it in and I tied on a better one—a white foam popper I had tied and given to Ted the last time I saw him. I rowed him into position and had him cast toward some lily pads. Bluegills charged his fly.

"God dammit, look at me! I'm fishing! I never thought it would happen again. Chris, take me out to one of those dark spots..."

How could I say no to this?

"Ted, I would love to, but I really have to get home. Let's head in and we'll do it for real on Monday!"

"Okay." Ted looked sullen, like a chastened child. I rowed us back toward the channel and helped Ted out of the boat, and we slowly made our way back to the truck, where he guided me through the reverse process of securing the boat to the bed.

We drove back, an hour or so later than I'd planned, and we settled on 5:00 a.m. for Monday. I told Ted that in the meantime I'd take his rod, rerig his leader and tippet, and tie on a better fly for the dawn hour.

So far this is a story about time management, and the lack thereof. But also about a becoming-friendship, and compromise: a negotiation over time. Again, fly-fishing has something to do with a peculiar inhabitance of temporality. It's about being out of joint with time but trying to create an impossible bridge across species and elements, in an absurdly elaborate fashion.

When I told my friend Mark about today, taking Ted out to test out our tactic for Monday, and about Ted himself, about to turn ninety-six and determined to go fishing one last time, Mark said, "You're going fishing with Father Time himself!" Mark didn't know that I was writing about time in this book, but it made me realize that there was more to this outing than just testing our route, more than just fishing. It was also about bridging generations, feeling the passing of time in a different way. I felt my forty-three-year-old body differently after my time with Ted. And my children looked different, too, when I returned home. Everything was stranger—and better for it.

There was an alpine lake in Montana that I hiked to once and camped by for a couple nights. A sheer cliff formed one side of the small body of water. It was surrounded by lodgepole pines and firs. And the lake was full of gregarious golden trout. It was more like a pond. I had brought my light rod and, casting from an outcropping, caught several fish and marveled at their brilliant coloration before releasing them back in the icy water. I wish I could recall the name of the lake, or the mountain, or the location of the trailhead. Instead, I have faint memories, a couple blurry photographs. Only the forever unrecoverable, the forgotten golden trout lake in my mind. And every past moment fishing ends up falling into this same abyss.

I was chatting with a local painter, David Grath, and he asked me what I was working on these days. I told him a book about fly-fishing, and he leaned on his cane and shifted into story mode: "When Jim Harrison taught me how to fly-fish, he told me that the rhythm of the cast was a figure 4"—he pantomimed a Platonic fly cast.

Casting has never felt like this to me. It's never been an aesthetic form or an abstract shape. It's a purely functional action, and I don't concentrate on what it looks like above me. I focus intently on the exact spot where I want my fly to land, and I put it there. That's it.

Fly casting takes practice, but it's not some discrete pattern or occult rhythm you're trying to master. It's just sticking with it, using the awkward technics of line and rod and arm and body and eyes to move the fly through the air some distance out and right where you want it to be. Where the fish are.

I was walking in the woods with Vera on my shoulders when I saw it: a barred owl, perched on the stump of a white ash. It was poised, its head turned just so I could see the slit of its left eye. We had been hearing the owls at dusk and had even spotted one swooping through the upper canopy a few days prior. But here it was, perfectly stationary a few feet above the ground, watching the forest floor: inevitably scanning for field mice or chipmunks. I whispered to Vera, "See it?" and she quietly exclaimed, "Yeah!" I crept closer, a foot at a time, tiptoeing across the oak leaves, and the owl, miraculously, stayed there still.

Only it didn't. It wasn't. An owl. It wasn't an owl. It was nothing more than the stump of the ash itself, spookily resembling the exact shape of an owl from a distance.

I've had this problem throughout my life: seeing things that definitely are not there. This has made me feel foolish many times, but it has also allowed me, I think, to sense when fish are present beneath the surface of the water. Because I visualize them, I imagine them—I conjure illusions of them that sometimes turn out to be or become real. They become fish. Thomas McGuane explains it like this: "The desire to see fish is a big part of the angler's enterprise" (273).

I was out with Glen once in the canoe and told him to cast his fly a foot to the right of a certain clump of reeds about fifty feet off our bow. "Right there?!"

"Right there."

Glen, being an expert caster, dropped the fly precisely in the spot I was pointing to, and the five-pound bass that I knew was waiting there engulfed it as soon as the fly touched the water. The bass, too, had been watching for something it knew would be there. A dragonfly—only it was Glen's fly instead. Glen eased the magnificent fish in and cradled it in the water in his net, marveling at the fish and marveling that it was *right there* where I'd said it would be. It had been nothing more than an illusion before it was real.

This happens to me nearly every time I go fly-fishing. I envision a fish where I know it should be, and it's there. It sounds preposterous, now that I'm writing it out. Or just all too obvious and common. People see things they want to see, mistake vague shapes for real entities all the time. Once, I was pouring balsamic vinegar into a dish of olive oil and saw a smiley face beaming up at me. But what I'm talking about with fish is different, somehow: it's seeing a form but also knowing what it will do. Knowing that if I set my fly down in a certain spot, the fish will zoom over from eight feet away, where I knew it was beneath an overhanging tree limb. And *seeing* it happen, across the often-indecipherable screen of the water's surface. Now this is just sounding like "reading the water," or strategic fishing. But when it actually happens, it feels much more mystical and uncanny, this anticipation of fish across our distinct realms.

A peculiar and wonderful thing about fishing dreams is how I can catch fish that do not exist, in bodies of water that flow only through my mind.

In early 2016 I wrote an essay for Jeffrey Jerome Cohen and Lowell Duckert's book *Veer Ecology*. This book was organized around a list of verbs that the authors "veered" with, to arrive at a surprising assortment of environmental epiphanies and ecological concepts. My chapter was called "Wait," and in it I attempted to bring together several strands of thought: some reflections on the temporality of waiting at airports, and what it means to wait in an ecological sense; a personal history of fishing; and a literary critical study of a few fishing scenes in fiction, and how they shed light on the ecological imperative to *wait*. This essay was one of my first attempts to write earnestly about fishing.

But by the time the book came out in 2017, the writing felt as if it took place in a different lifetime. In the strange years of the Donald Trump presidency, I didn't feel quite so sanguine about pondering this wait. It didn't seem right to urge people to *wait*. I didn't realize that "staying with the trouble" would be this difficult, or take this exact form. Nevertheless, there we were for four years.

On sabbatical during the 2016–17 academic year, I was fly-fishing more than usual. Part of this was planned, but increasingly I fly-fished to block out the noise of the political gyre whirling around. Standing waist-deep in water, tuned in to myriad life forms and elements around me, casting at a ripple over here, the edge of a cluster of water lilies over there—this was a regular sort of escape for me over that turbulent year. Escaping the urgency of the moment, if only for a few hours at a time, in order to wait for other things to happen.

Sometime during that year, a friend who knows I like to fly-fish forwarded a *Wall Street Journal* article called—wait for it—"How the Rich Fish." It began: "Some avid anglers travel thousands of miles to fly-fish for trout in the rivers and streams of the Rocky Mountains." I wrote about this type of fishing person in my *Veer Ecology* chapter, how I recall them becoming impatient and angry in the Bozeman airport (where I worked for a couple years) after their idyllic trips when faced with a logistical obstacle out of their control. On the river it's one thing to deal with exigencies: to wait for the perfect trout, or for the wind to suddenly cease, or for the right moment to set the hook—this is part of the adventure. But in the airport? Interruptions and delays are just a nuisance, annoying obligations to wait.

The *Wall Street Journal* continued by contrasting these sorry, wealthy travelers with another kind of angler, one who doesn't have to wait around for fish: "Don Felsinger can cast a line from his patio. Over 2 acres of man-made ponds and streams stocked with native cutthroat trout surround the 8,000-square-foot-home that Mr. Felsinger and his wife, Jenny, built on 15 acres in Jackson Hole, Wyo., in 2015." Consider this veritable Xanadu, with its "man-made" ponds and streams, and its colossal home. The fish are at once stocked and native: a finessed contradiction, indeed.

In many ways it's a typical *Wall Street Journal* article, celebrating exorbitant luxury as a natural state of affairs, and an obviously desirable destination. The article goes on to explain, "To lure home buyers hooked on fly fishing, developers are creating luxury ranch resorts that offer blue-ribbon trout streams along with exclusive amenities. Some anglers are even designing their own private fisheries, with help from a cadre of biologists, stream restoration specialists and water-rights consultants that has grown to meet the rising demand for luxury angling properties."

As galling as this classist terraforming endeavor sounds on one level, we also get a bizarre glimpse into something else, something almost resembling *ecology*: a "cadre of biologists," "stream restoration," "water-rights" . . . While it is understood that these things are strictly in the service of property owners who literally *value* these things and attendant activities (rivers, trout, fishing), there is still a curious, if vexing, form of conscientious coexistence happening here:

> At Mr. Felsinger's home, four computer-controlled pumps regulate the temperature and flow rate of his ponds and recirculating streams. A redwood water wheel made by their

home builder . . . aerates the water; on very hot days, an air compressor pumps in extra oxygen. Undercut banks planted with wetland grasses shelter the trout from eagles and attract tasty bugs. Gravel beds provide a spawning habitat during the spring runoff season, when rivers and streams in the Rockies swell with snowmelt—or in this case, when Mr. Felsinger taps a touch-screen panel to release a gush of cold groundwater from a dedicated well.

I can't shake this article, not because it makes me want a "man-made" pond of my own, but because it veers me into the realm of ecological thought. Yet aside from rigorous (if perhaps unsurprising) class analysis, or just vehement disavowal, I'm not sure what to do with these curated trout streams, this cyborg portrait of a pastime I love and loath, by turns and sometimes at once.

I know that my own interest in fly-fishing—which I'm at pains to say is very different from the sort on display in the *Wall Street Journal* article—is deeply concerned with matters of coexistence and the requisite lingering that takes place among and between species, landforms, and the complexities of planetary being. My version of fly-fishing echoes that of Jan Zita Grover, who in her book *Northern Waters* writes, "Submitting myself to the tutelage of water, weather, and fish doesn't lead only inward; it also leads toward the immense, seemingly intractable problems of logging, sewage disposal, and mercury deposition— problems that no one can escape" (99). I said earlier that fly-fishing has been an escape for me, but it's really been more of a watery grounding, a way to inhabit and live through this excruciating wait.

Fly-fishing is active, but it's also about *waiting*—about knowing how to wait and yet also not knowing for how long, or what for, exactly. And most of the time, the something else never arrives.

Fishing memories are slippery things. In the epilogue of *On the Water*, Guy de la Valdène notes how "remembering the details of one's life decades after the fact is not a simple matter" (180). The water and the otherness of fish compound this problem.

Sometimes people kiss fish before releasing them. I can't remember where I first saw this ritual, perhaps in a magazine, or a fishing show. I think I have done it myself a few times, possibly more, mimicking the bass masters I admired. What is this sign of intimacy? Jan Zita Grover gets at it in *Northern Waters*, reflecting on the need "to touch fish with wet and, I hope, reverent hands, to feel their bodies torquing away from me, leaving a fresh and acrid scent on my fingers that lingers as potently, as alluringly, as the smell of sex" (5). I know this feeling. The slime of the fish, the long-lasting and pungent fragrance, the mortal struggle and passion intertwined . . .

One of my early fishing mentors was named Steve. He lived next to a cemetery near one of the lakes in the park; he was a custodian of the property, or something like that. He lived in a trailer with his wife, and we met near the lake once and he invited me fishing on his twelve-foot aluminum rowboat, with his father and younger brother. It seems doubtful that all four of us fit in that tiny craft, but my memory has us all in there, casting top-water baits toward the pondweed stalks at dusk. I was in high school at this time.

Usually Steve would call me up in the evenings and invite me to go fishing, then pick me up, and we would fish until sundown. I learned a lot about the lakes during those outings, their drop-offs and sandbars and submerged points. I also learned how to handle a rowboat—not a terribly complicated thing, but a minor art at least.

Once, as we were fishing, Steve told me apropos of nothing that he liked "to watch a *leeeetle* bit of the Playboy channel before going to bed." I remember being caught off guard by this statement, and I just fished through it, focusing on some water lotuses that looked particularly promising for giving cover to a big bass. To this day I don't know why Steve brought that up, or where he expected the comment to go. But there we were fishing together, trying to connect with these other species across elements, suddenly talking indirectly about sex. Or, really, not about sex at all: about *not* having sex, apparently. About displaced, perhaps repressed desire. As for me, I hadn't even discovered masturbation at that point. My parents moved our family around a lot before I started high school, and so romantic connections or even fantasies never quite had time to take hold. It's possible, looking back, that fishing served as a constant where I didn't have others.

The following fall, after a busy summer of washing dishes for a local summer camp, I called Steve to see if he wanted to go fishing. He told me he and his wife had divorced, and he no longer lived by the graveyard in the park. He didn't get back over to the lakes as often, but we'd try to make another time work. I never fished with Steve again.

Two days to go, before my fishing trip with Ted.

As I mentioned at the beginning of this book, I didn't fish at all during the four years I was in college, but after I graduated and moved to Wyoming, where I was guiding river trips for a summer job, a couple of my fellow guides took me along one evening to the Lewis River in Yellowstone, where I promptly caught several brown trout—and I fell back in love with fishing. I spent a chunk of my tip money on a five-weight Orvis fly rod and reel, and a floating line, and I cobbled together a box of dry flies that never failed to catch trout—mostly Adams and Wulff patterns. From that point on, every evening, after we'd put away the rafts, I'd head down to the Snake River and fish for cutthroats, watching the surface of the water come to life with insects and feeding trout.

Living in Bozeman, Montana, over the following few years as I worked on my master's degree, I got to know parts of the Gallatin, Madison, and Jefferson Rivers with the help of the best fishing partner one could ever ask for, Greg Keeler. Greg taught creative writing in the program I was enrolled in, and was something of a local legend. I introduced myself to him one day and mentioned I, too, loved to fish, and we quickly realized that we were kindred spirits: fishing fanatics of the same ilk. Before long we were headed out to his trusty swirly holes on those main rivers, the three forks of the Missouri—as well as to some more obscure creeks and lakes in the region—several times a week. Our routine consisted of a phone call, a meetup at his house shortly thereafter (I lived in an apartment just a couple blocks from his house, as it turned out), a quick stop at McDonald's on our way out of town—Big N' Tasty meals were our favorites—and then the breakneck-speed drive along Highway 90 toward one of our spots. The reason we always ate at McDonald's first, I think, had to do with the fact that I was writing a series of postmodern stories for the fiction class I was taking with Greg at the time; they were college stories framed around adventures I had with my roommate and best friend, Tyler, and every story had a fast-food restaurant scene, which I described in painstaking detail. Greg either was indulging the consumer scenery I was writing about or just liked McDonald's, I was never quite sure. I didn't eat fast food that much those days, but Greg always treated, and I always enjoyed the meals on our way to fish.

One of Greg's favorite flies, one he invented, was called a "Bugger Duck." Born out of equal parts satire and necessity, it is a fly made from scraps of fishing supplies salvaged in the back of a pickup truck, culminating in a strip of duct tape around a found dead duck's feathers. It looks awful and is anathema to fly-fishing purists, and yet it catches fish. This was one of Greg's performative acts of resistance to the rise in fly-fishing tourism in the American West during the later decades of the twentieth century. He watched the rivers in Montana turn into carnivalesque aquatic superhighways of colorful boats, angrily threshing clients, and shouting cynical guides. Still, he knew where we could go to find quieter fishing spots. The rivers still held many secrets.

Looking back, I can't believe I waded in such unfamiliar waters. Those rivers are so strong! But Greg knew them well enough that he would simply wade in and I would follow, and he would lead us to hidden cutbanks, irrigation dams, and boulder fields where we would inevitably find hungry trout. Greg would hand me the right nymph or tell me that one I had already would work perfectly, and while I was not a skilled caster, he never lectured or gave me pointers but just let me flail around and, incredibly, catch trout. I can conjure many of the bends of the Gallatin and Madison that we would fish—although I'm not sure I could find my way back to those places. This is when we would fly-fish, which wasn't the only way we fished.

Sucker chunks: this was perhaps Greg's preferred mode of fishing, and it is the antithesis of fly-fishing. But it's what we did first, before we ever fly-fished together. When Greg first called me on the phone, he asked me if I had a spinning rod—I didn't at that point; all my old gear was up in Michigan in my parents' basement. I only had my Orvis fly rod there. He said no matter, he had an extra. I was confused and didn't know what we'd be doing, but I was up for it. After we hit McDonald's and the high-fructose corn syrup injection kicked in, we drove thirty minutes or so to the spot: a dodgy dirt road next to some railroad tracks with NO ENTRY and DO NOT TRESPASS signs festooning the barbed wire we climbed through. I remember feeling nervous. Greg was holding two spinning rods and had what I would later learn he called his "death bag" over his shoulder: an old leather satchel that he kept his miscellaneous gear in—and trout, when he caught them. When we got to a gravelly point that dissolved into the churning river—this was near the headwaters of the Missouri—Greg rigged up the two rods with gobs of nightcrawlers and heavy sinkers about a foot up from the hook. This was like my first fishing, when I was ten or so. What were we doing here, the mountains sweeping the horizon in the distance and cottonwoods snickering in the breeze? Greg showed me how to lob the massive bait+sinker assemblage out toward the edge of the current and then follow it as it bumped across the bottom and nested into place seventy-five degrees downstream. Then, to wait for the tick-tick-tick that signaled action. But this would not come immediately. No, we placed the rods on the ground, angled the tips just so, and put large rocks over the rod butts to hold them in place. Then we stood there with our hands in our pockets, waiting. When the tick-tick-tick happened, Greg would raise the rod gently and then give it a good yank—and then reel in the fish: a sucker. These are bottom-feeding fish that are not exactly "sport" fish, but for Greg they were a part of a larger process. Having caught a couple suckers, Greg would unfold his fishing knife and carve the sucker fillets into one-inch-square chunks of meat: "belly bait." He would put these pieces of bait in Ziploc bags (one for us today, another for his freezer, for our next outing), and we would then drive to *another* curve of the river, upstream. There we would use the same hooks and sinkers, but with belly bait instead of worms. And we were casting into a tighter pool, an eddy where the current wrapped around the bend. Here we were after brown and rainbow trout, and we would usually catch a few nice trout and take them back to town, where Greg would gut them and then put them in his homemade smokehouse: an old refrigerator with a chimney pipe installed, stationed out in the alley behind his house. A perpetually smoldering fire was down where the freezer had been. This was not fly-

Fly-Fishing   63

fishing, but Greg's smoked trout were delicious, the best I've ever tasted.

Yet Greg taught me enough about fly-fishing in the area, too, that I got to know some choice stretches of the rivers in the valleys that surround Bozeman. We also hunted for morels and wild asparagus by the rivers, and sometimes our outings would switch from fishing to foraging in an instant, if we saw a different kind of bounty.

One time a fellow graduate student who idolized *A River Runs through It* discovered I liked to fly-fish, and he asked me to take him out, so I did, letting him borrow a backup rod I had since obtained. We went out to a part of the Gallatin that was easy to wade in sandals, and it was just warm enough to do so without your legs freezing stiff—it was early June. I showed Ben how to do a simple roll cast into the current, and how to target the pools behind boulders. It was a pristine Big Sky day, but the fishing was slow. I caught a couple small rainbows, then left Ben practicing his cast while I made my way downstream. Near a stand of cottonwoods I saw them: a cluster of perfect morels. I stood my rod against the trunk of one of the trees and proceeded to fill up my hat with the mushrooms. I heard Ben shouting—he'd lost sight of me—and I dropped back down onto the bank, grinning and holding up my treasure. "Check it out, morels!"

Ben looked glum and remarked tersely, "I thought we were here to fish."

I halfheartedly cast a few more times and tried to get Ben into a place where he might hook a trout, but you can't force these things. Ben was losing interest, and I was so distracted by the fungus trove that I wasn't intent on catching fish anymore. We drove back to town in near silence, our one and only fishing excursion together. Later, Ben wrote a good chapter of a book on Brad Pitt that I coedited, about Montana and Pitt's depiction of Paul in the film adaptation of *A River Runs through It*. So that fishing trip worked out, in the end.

While I was writing this book, Greg sent me the sonnet that serves as the epigraph, and I asked him if I could include it here. At first I thought I might write about this poem, "analyze" it or "unpack" it, but that ended up feeling silly. It's pretty direct, playful ambiguity and all. So I just decided to make it the beginning of the book. It sets the right tone: it ties together fishing and writing, which is something I've been thinking about as I've written this book, remembered fishing times earlier in my life, and gone fishing between writing sessions.

For over twenty years Greg has been writing a sonnet every morning—I know, because he sends them to me around the same time each day. Once when Greg visited me in Michigan, I was taking him out fly-fishing on my lakes so much and so early each day that he missed a couple days of his sonnets. I don't think it really bugged him all that much, but I noticed—and so did the other bunch of people whom he sends them to every morning.

Greg wrote a book once called *Trash Fish*, his own entry in this choked genre of fishing books. It records his life in fishing and his midlife crisis and how things gradually stabilized around and within him. I have a cameo appearance toward the end of the book, as we met during the time he was writing it. Unknowingly, I was also his sidekick as he was writing another book about his friend Richard Brautigan; we would take these long detours on our way to fish, and suddenly Greg would pull over and stare at a dilapidated barn or a shot-up berm. Then he'd drive away, and we'd eventually find our way back to the river and be fishing again. I didn't know it at the time, but Greg was revisiting old haunts where he and Richard had gotten into one sort of trouble or another—places he hadn't been in a while, after Richard's suicide. He was working on his book *Waltzing with the Captain*, in an aslant way. I would find out about this only later when he gave me a copy of the book and I recognized a lot of the locations from circuitous routes we'd taken on our way to fishing.

After I finished my master's degree at Montana State University, I moved on to Davis, California, to pursue my PhD, and once again fishing took a backseat to other things, in particular learning to teach and learning (always learning) to read and write. My rod case was tucked away in my closet for six years as I taught literature and writing, and as I read copious books and articles for my seminars, and as I wrote endless papers and eventually a three-hundred-page dissertation about airports in American literature—partly inspired by my time in Bozeman, where I worked at the airport outside of town. Greg and I stayed in touch throughout that time, I visited him when I could, and we continue to email every day and meet to fish every couple of years—either up in Michigan or in Bozeman, when time permits.

After a good fishing day, as I fade off to sleep I will see fish exploding out of the water on the horizon of my consciousness. Fly-fishing takes place on various uncertain thresholds: between surface and submarine, between species, between artifice and organic life. This time before sleep, drifting into unconsciousness, is another such threshold—a state that might be curiously closer to the tactical nature of fly-fishing than it might at first appear.

During the spring of 2021, Vera and I had a midmorning regimen: after taking Camille to school (Julien was still home doing school on Zoom), we would take a quick stroller walk down to the end of Bayou St. John and back home. I would pack my small tenkara rod and a box full of flies. If we saw some bass cruising, or cichlids, or bluegills, or other mystery fish, I'd cast for a few minutes—often catching one fish, to Vera's delight. I'd let her touch the fin, look into the fish's eyes, and then we'd release it.

I had tied some very small minnows out of belly-white EP fibers and red Crystal Flash—just the barest suggestion of an injured creature. I wanted to see how they swam in the tepid water, and we had seen cichlids holding near a concrete drainage housing. But when we got there that morning, the cichlids were gone. Fish appear and disappear, often without our understanding why.

But I still wanted to see how my little fly looked in the water, so I cast it toward the concrete square and immediately saw a flash and felt a strong tug on the line—I missed this fish but put the fly right back out into the middle of the ripple. Another flash, and this time I hooked it: it made a full bow in my rod, and as I brought it in, cars stopped on the street a few feet away to watch. I pulled up a glowing dark bluegill, plump and maybe five inches—a glorious prize. I showed to Vera and set it down in the sludgy water, watching it kick back into the near depths. I cast again to the same place, and *bam*, another strike—another strong bluegill. I caught three in a row, each one more colorful and vibrant than the last (or maybe it always just feels that way, living in time). I missed a few, too—so it was definitely a pod of bluegills, staging or feeding prespawn.

Vera was getting antsy, and so I collapsed the rod and we started our trek home, passing white herons in the willows and crows on the light poles. The next day I went back and tried to repeat the action, but there were no signs of the fish. I cast futilely toward the same oblique spot where the bluegills had been, but it was a one-time thing, a fluke—and maybe all the better for it. Those sudden flashes under the surface still play before my mind's eye today.

When I fish at the cement-lined end of Bayou St. John, I sometimes see an enormous golden koi swimming lazily along the bottom. It appears in the murk in front of me for a few moments and then disappears again. It must be three feet long and almost a foot wide. Someone released it as a small goldfish, perhaps, from a fish tank many years ago.

I've always wanted to catch a koi, or carp, on a fly rod. Carp, or *Cyprinus carpio*, have become a new kind of sportfish up in Michigan. In his book *Fly Fishing the Inland Oceans*, Jerry Darkes describes how carp will "test equipment to the limit" (38)—a wonderfully material way to explain the payoff of fishing for these creatures that, as Darkes also notes, have been "worshipped and revered" (35) in parts of the world. They are powerful and mystical.

I've tried a few times to fish for them, casting into the pods of roving carp that come onto the shallow flats of Lake Michigan to feed in the summer, trying to land a small crustacean fly far enough in front of their searching barbels, anticipating their path so as not to spook them. But I have not succeeded in hooking one, or even generating anything close to interest from them. At least not that I have been able to tell. But I haven't really spent time focusing on this. Maybe next summer.

Yet watching them alone is something, their giant shimmering bodies moving across the sand like small plump sharks. *Carp* makes them sound grudging, or annoying—when, in fact, they are truly majestic fish, sensitive and strong and serene.

I did fish for koi during several months of my life, amassing a colorful collection of various fish of different sizes and patterns, which I kept in multiple ornamental ponds. This took place on my first iPad, back in 2012, on an app called Koi Pond. Sadly it was not fly-fishing on that app but a crude simulacrum of bait fishing with a bobber. You "paid" for more expensive bait, which then nonsensically resulted in rarer (and more valuable, in the game's currency) koi. The goal of the game, such as there was one, was to accumulate more and more ponds full of more and more koi, which accrued value as they grew over time, if you fed them regularly and kept the ponds clean—in other words, maintained the app each day. Koi could be grown more quickly by exchanging in-app "koins," then sold back to the game's fictional store to get more koins. Meanwhile, you could festoon the ponds with decorative plants and fill them with other creatures, such as frogs, turtles, and dragonflies—all of which cost koins as well. The game was, in other words, a rectangular Zenified microcosm of capitalism, ported through the iPad's app.

I was wasting time with this app during a period when I was not fly-fishing. The most recent fishing I'd done at that point was during a brief canoe paddle around one of my favorite lakes with my father-in-law, Peter, in July 2010, just before the birth of Julien. I used some of my old gear from my adolescence, still very much functional: a medium-action spinning rod, some ten-pound test line, a size 2/0 hook, and a six-inch-long "Jerk-worm" that mimics a darting minnow beneath the surface. I caught several bass that evening, surprising Peter with my crystal-clear memories from fifteen years prior of exactly where the fish would be.

Two summers later I would take Peter out in the canoe again for his fifty-seventh birthday—we went to the lake of my childhood map. I didn't fish that time; we spotted herons and kingfishers along the shore, and at one point watched in awe as ten sandhill cranes soared directly over us.

Late that fall, Peter started dying of cancer. We visited him in January of the new year, in the Christian Science care facility he had chosen to stay in—an expansive resort-like campus, built in the early 1950s in anticipation of countless followers of Christian Science who, in lieu of medical treatment, were supposed to enter for brief stays and then return to society after spiritual healing. Now, the population of that minor religion having dwindled, the place was practically empty—long, dark halls with only occasional muffled moans emanating from a few closed doors. The campus is located on a barren street called Rott Road.

During that visit Peter was fiddling with a still-new iPhone that he had gotten right before admitting himself to the facility.

I downloaded the Koi Pond app for him and showed him how to play. He seemed relieved to have a modicum of distraction as he lay there in pain, and he started fishing for koi and collecting them in his pond. A little scene of tranquility in a horror show.

In March we got the phone call to fly back up to St. Louis to say goodbye to Peter. The agony having crossed a threshold, Peter had made the call and had been extracted by EMTs and taken by an ambulance to the hospital, where he had last-resort chemotherapy but, more important, was given pain medication. At one point before Lara and my mother-in-law, Marcia, went to visit Peter at the hospital, Marcia handed me Peter's iPhone. "Here, take it . . . "

I thumbed the phone's screen and saw the last call made: *911*. I toggled over to his Koi Pond app and watched as several white bellies floated in the green murk of Peter's pond. That's what happened when you stopped using the app: the koi died, and the ponds went stagnant. Peter died two days later.

I'm looking back at the map of my favorite lake I made when I was young. What was I doing with this map? Trying to understand this small body of water, trying to learn patterns about the various species of fish that inhabited it. It was my "favorite" lake owing to proximity: it was five miles from our home, and I could be dropped off by my parents and picked up later without too much hassle. I mapped it because I went there so much. It wasn't about mapping the unknown, more about getting to know it better—to better appreciate its mysteries, even.

But what else? I decided to ask my childhood friend Jon Bonkoski, who now works as a Geographic Information Systems analyst for an ecological firm out of Portland, Oregon; Jon specializes in Pacific coast salmon fisheries, especially up in Alaska. Jon and I have never fished together, but he knows my obsession and has canoed some of the lakes with me up in Michigan. We originally met at summer camp when we were around ten. I emailed Jon these questions:

> *What do maps reveal about fisheries or fish ecosystems that humans otherwise miss or misperceive? But also: In what ways have maps misguided or muddled human knowledge of fish ecosystems? (Maybe this is more historical, as in, less accurate maps.)*

Jon replied:

> *Interesting questions—I've been contemplating a similar set recently because of a spatial analysis I've been doing on older fisheries datasets from some earlier work I did in California. From a broad perspective, maps are a tricky thing. People look at maps and take them as gospel—this is where things are and how they look—which is not true in the slightest. To begin with, maps are flat and the world is round so maps are approximations at best. We use fairly sophisticated math to project a curved area onto a flat surface but invariably there are distortions. The Catholic Church created maps using the Mercator projection that made their reach seem endless. The other aspect of mapping that people don't typically consider is that maps are temporal. Maps are akin to a photograph—a snapshot of the past. By the time a map (or photograph) is printed that time has passed and the map or picture is a depiction of something that was*

and not necessarily a current reflection of reality. Bringing this around to fisheries mapping, both of these concepts are true—fisheries maps are approximations of past knowledge. Furthermore, when mapping fisheries you're not mapping fish at all, you're mapping fish habitat. Fish like a lot of animals move around, so when you map fisheries you're mapping a set of conditions that meet certain species' lifecycle needs. For example, a major fishery on the west coast is the groundfish fishery. This is a set of 200+ species, like rockfish and others, that are essentially non-migratory— that is, they stick around piles of rocks and corals. When you map these species, you're mapping the piles of rocks and coral. The resulting map is a depiction of areas that meet the habitat criteria for a species but there is no guarantee fish will be there. And, when you consider highly-migratory species like salmon or tuna—what are the criteria you're mapping? How do you map a species that ranges the entire North Pacific? And what do you gain from making that map? I guess the bigger point is that fisheries maps are ultimately a depiction of where humans interface with fish. It's where we catch fish or hope to catch fish. When we create Marine Protected Areas, we are not managing fisheries— we are managing people. We are reducing the instances of human interaction.

A snapshot of the past. I love that phrase. I didn't expect Jon to go into this much detail, but his observation at the end gets perfectly at the dehumanizing inversion I have been trying to think about as I've written this book. I am not fly-fishing to *master* the fish or the water but to actually experience a temporary "interface," to use Jon's word, between me and all the other entities and ecological formations that encompass us. My crude lake maps were in their rough way an attempt to catalog snapshots of interfaces in this habitat, modest records of extremely temporary encounters across always shifting ground. Even when I look at the lakes I fish in Michigan from satellite views, I am always struck by the disconnect between the visible contours and the real textures of these places.

Tomorrow morning at 5:00 a.m., I am to meet Ted in our driveway and set out for our fishing trip. It's a calm night tonight, the sun sinking toward Lake Michigan, into a rosé layer of western clouds. Any other evening like this, I'd be champing at the bit to go to sleep so I could wake up and fish. But tonight I am sitting here writing—preparing for the morning to come. What will happen? Will Ted show up on time? (Most likely, yes; Ted was a B-17 turret gunner who flew thirty missions over Germany in the 1940s. I don't think that punctuality is one of his problems.) How will the fishing be? I hope I can get Ted to the best spots on the lake, where I am confident he will catch fish. I have his rod all ready to go; I've never held such a nice rod and reel. For myself, I am taking my small, relatively cheap three-weight outfit—in part because I want to be mostly attentive to Ted, and I'll be less tempted to make long casts with big flies toward lurking giant bass if I am restricted to my lightest rod. Instead, I'll constrain my fishing tomorrow to the bluegills and sunfish, closer to the boat. (I'm also a little paranoid that we will flip the boat at some point, and all will be lost—and I'd rather not risk forfeiting one of my nicer outfits.) How long will Ted want to stay out on the water? Will he get noticeably tired? What will be my cue to head back in? Do I offer to drive this time? What if we crash into a deer or, worse, a tree on our way to the lake? So much for excitement the night before fishing. Instead, I am doomed to wait it out until 4:44, when I can turn off the alarm on my phone before it has a chance to go off. This is not the kind of waiting that I typically deal with when it comes to fly-fishing.

There's a lake I fish a couple times a year that is haunted. The local conservancy describes a trail around the lake as on "land once used for Ottawa and Chippewa settlements." This puts it too neatly. I feel *unsettled* when I'm at this lake. It has at least something to do with the heavy, still-unanswered history of settler colonialism and the eviction of Indigenous peoples in this region and on this continent (not to mention around the world). Peshawbestown is a small village nearby, part of a larger Native American reservation in the area dating from the mid-1800s. This history is present here.

The lake is small and can be waded in its entirety, but it confounds me each time I go there. When I look at it on Google Earth, its apparent straightforwardness mocks me—it's deeper and weirder than it looks from the all-knowing above. To make sure I wasn't off in my assessment of this lake, I took Glen there a few years ago, and he agreed: "That lake is scary." While we each caught several formidable fish there, it's as if the lake gives some gifts but then withholds much, much more. It's a difficult feeling to explain: a weight, a guilt—a vivid unpaid debt. Glen and I tune into it when we're here, and it won't let us alone. And I am not complaining, because these feelings should be more palpable more often, across this continent.

The water is full of the same fish that I pursue in the other lakes—bass, bluegill, and the occasional pike—but the fish are fewer and farther between. The lake is surrounded by ferns on the shore, and reeds and sedges and arrowhead in the shallows. Pondweed walls encircle the shoreline, about thirty feet out. There is a spring-fed marsh on one end, full of lily pads and bladderwort. Loons roost on this lake, and once when I was fishing there a coyote loped along the shoreline, eyeing me as it passed. When the fish are active, it can be very exciting. But many times at this lake it is uncomfortably quiet and atmospherically enigmatic. It feels like I could be somewhere else far away in space or time. The air above the lake pushes down, amplifying sounds and reverberations. I hear each pileated woodpecker tap, each rustle of the hemlock needles as the trees sway. It's almost like I've stumbled into Ursula Le Guin's alternate ecology of *The Beginning Place*.

There's one shoreline whose shallows are covered with strange light-colored stones the size of dinner plates, objects that my friend KT Thompson might call, echoing Willa Cather in *Death Comes for the Archbishop*, "queer rocks." These are geological fragments that signify inaccessible pasts—but also, perhaps, a troubled present. They are the substrate beneath where violence has taken place, and anxious reproduction at the expense of earlier peoples.

Walking delicately over these stones, unsure of their history and meaning, I feel the uneasiness of this place pressing up from below. It's a strange lake, and even as I cast toward banks of reeds and occasionally catch a fish, I am conscious of something *off* here. I'll return to this lake, but with apprehension. And, in truth, isn't this all the lakes I fish? The tectonic plates on this continent are all queer rocks, as we live through unfinished—and tarnished—history.

I said that I didn't fish at all when I was in college, but now that I'm piecing back together the tapestry of this practice in my life, I realize my timeline must be skewed: that first summer in Wyoming was actually after my junior year of college, and I must have had at least a brush with fly-fishing. Because somehow I ended up writing letters to my college friend Jennifer about fly-fishing; she was from Sun Valley, Idaho, and grew up fly-fishing for trout around her home region. She was back there for the summer, and we made vague plans to go fly-fishing together. Her father was a pilot and had a Cessna; we could meet at the Jackson airport, and he'd fly me over for a few days.

It didn't happen, though, that time. Our lives took separate courses, and we never met and fished together in Idaho. But every time I meet someone from Sun Valley, or read about the place—Mark Kurlansky lingers there, tantalizingly, in his new book—I think about Jennifer and wonder what it would have been like to fish those rivers and streams with her.

This morning is thick with bird songs and the heaviness of sky—I can smell the rain coming. It would be a perfect morning to slip out and head to one of the lakes, but instead I am back here in front of my computer, sitting at the kitchen table, watching the clouds amass above Lake Michigan, a pot of black tea next to me. I won't go out this morning because yesterday was Monday, when I finally went out with Ted. I need a day to recover and recalibrate. Let me back up thirty-six hours.

I did not sleep much the night before. I was nervous about all the factors and anxious to just get it over with. Not the fishing, of course, but the *guiding*. Ted was keenly aware of not wanting me to *feel* like a guide, but he knew this was basically the reality: he needed help. I packed my backpack with a bailer (the boat was maybe taking on water when we did our test run—though Ted swore it wasn't), my improvised anchor made out of a horseshoe and a thin but strong cord (Ted's own improvised anchor consisted of a half-gallon plastic milk bottle that had been filled with cement ages ago—it was coming apart and unwieldly, and the rope it was tied to was a mess), and my life jacket—I wanted to make sure we had life jackets, and I hadn't seen any in Ted's truck.

I thrashed around all night, all the great anticipation of fishing having turned into existential mush. I had hoped to wake up right before my alarm sounded so I could, as usual, shut it off before it woke up Lara or anyone else. But I was fully awake at 2:57, and I realized I was just going to have to wait it out. I figured if I went through every step in my mind, mentally taking myself through what would (or might) happen, I could consume the time and get myself there. I began to go through it all in my head, from getting out of bed to walking down the driveway to where we would meet, and so on. I had gotten pretty far into the morning when the alarm blared and I woke with a start—I'd fallen dead asleep, of course.

Flustered, with my heart racing, because that's what alarms do, I dressed, gathered my pack and our rods, and walked down in the darkness to meet Ted.

I was meeting Ted down the drive at my parents' house, where deer had been decimating the gardens recently. When I arrived, I heard a loud snort, and a few deer stomped off into the woods. This was *their* time, being crepuscular creatures. I stood near the vegetable garden waiting. 4:57. The deer reapproached, until they detected me and again stormed off, making angry sounds and crashing through the undergrowth—back and forth, to really make a point of it. It got so loud at one point that I thought, *Great, before Ted can even arrive, I'm going to be gored by a raging deer. Just great.*

I looked up at the brilliant stars, watching for the eerily bright strings of Starlink satellites that locals had been complaining about. Elon Musk's stars, or, more lately, Richard Branson's. Next week Jeff Bezos was to launch off. A profound waste, this refurbished spectacle of the past century's excesses. Say what you will about space exploration, but there will be no fly-fishing on Mars or anywhere else in the nearby galaxy.

At 5:01 I heard Ted's F350 roar to life across the hills, and I tracked the sound of his truck as he drove around the corner and up the driveway. Here he was, just a few minutes late—apologizing profusely when I opened the door. I was just glad to be, finally, on our way. "It's fine! Here we go, Ted!"

As we pulled out into the road, another vehicle was approaching from the south. Ted fumbled to find the switch to turn down the high beams—and managed to turn the headlights off completely. So we raced through the dark toward the approaching car, as Ted continued to flip switches until miraculously the headlights came back on. Our first crisis averted.

We reached our spot without further trouble, and I hopped out and began the baroque process of getting the boat unloaded and ready. I removed all the old junk from the bottom of the boat, took out the oars, and set everything on the ground near the truck, and then Ted and I slid the boat back and down. I eased up the hull and slid the wheels beneath, and then bungeed it all together as Ted had instructed me. (I had brought plenty of simple climbing cord as a backup; I do not trust these old rubber-and-metal-hook death cables.) I put everything back in the boat, and Ted said, "Hang on, let me just get my life jacket; I always wear a life jacket." Hooray! Ted dug through the space behind his seat and pulled out a natty green vest. "Chris, this life jacket is fifty years old. I'm not sure it even floats." *Well, it's the thought that counts*. To top it off: he didn't even zip up the vest, just let it hang off his shoulders. If he fell in the water, he was going to sink like a stone.

We started our trek to the lake. The idea was to be on the water by sunrise at 6. So far, so good.

At the shoreline I removed the wheels and set the rods in place so they wouldn't get snapped or hung up on trees as we shoved off. Ted crawled into the bow and got himself situated; I pushed the boat out into the shallows, waded back and climbed into the stern, and used a canoe paddle that Ted had brought to scull us through the weeds and toward the open water.

I haven't mentioned it yet, I think because I'm still holding my breath in my memory, but the conditions were superb: it was utterly still and deliciously misty over the lake.

Out just past the sedges, I told Ted to cast toward a clump of tall reeds twenty feet away. He pulled out enough line and cast, getting the feel of the slightly larger fly I'd tied on for him. The fly landed right where it should and almost immediately disappeared. "*Ted, you got one!*"

He pulled up and had a nice bass on. "Oh! *This* is what I came for!" Ted started to bring in the line, but the bass beelined for the boat and shook off before Ted could land it. But it was enough. It's 5:56 when this happened—I know because I snapped a photo of it right before the bass got free, and can see the timestamp on my phone.

The next three hours were a blur of catching fish, missing fish (a lot of missed fish), and generally doing what it was we came for, to use Ted's phrase. We cast our flies toward lily pads and submerged cedar trunks and dogwood branches and reeds. We drifted up onto one pod of giant bluegills and caught about ten in a row, including the largest bluegill Ted had ever caught. He was ecstatic. Catching a bluegill on a light fly rod is one of the most exhilarating experiences, similar, I imagine, to the thrill that saltwater anglers get from the massive broad-body fish like permit—just scaled down.

I was using my very cheap three-weight rod, an Eagle Claw fiberglass rod I bought for $30, intending originally to use it to teach Julien how to fly cast. It seemed like the perfect entry rod, and I wouldn't worry about it snapping or getting scuffed up. But the first time I used it, it was so much fun that I kept taking it out—it flexes so supplely and casts surprisingly well. But I wasn't ready for the six-pound bass that slurped my fly at one point, pulling Ted's boat and charging into the reeds. I managed to land the bass—my biggest of the season and an unexpected reward for this trip, an unlikely catch on my petite three-weight.

We went up the shoreline fishing along the way, but by 9:15 I could tell Ted was getting uncomfortable—he couldn't find a good position to be in. I asked him if he wanted to head back in, but he was determined to keep going—there was a specific spot he wanted to show me at the other end of the lake. So we made our way, but at one point Ted tried to move from the front seat to the middle seat, so as to stretch his legs, but as he settled into position, I heard a terrible *crack* and Ted was down—lying across the floor of the boat, his neck wedged against my seat. The middle seat's rear thwart had snapped. For a moment he didn't move, and I was sure he was dead. No, he was fine, he said—just really pissed off, mostly because he'd broken his boat.

Luckily we were in shallow water, and I quickly swung my legs over the gunnel and stepped into the water so I could assess the situation and help Ted. He didn't look okay, but he said he was okay and, laughing, said it actually felt good to be on his back for a change. He said, "Oh Chris, I bet you weren't counting on this today." I thought to myself, *Actually, I was counting on exactly this. I just didn't know when it would happen or how bad it would be.*

I made sure Ted was actually okay, and he seemed to be, and I asked how I could help. He wanted to rest there for a few minutes, and then we'd try to get him back on the seat. Slowly and carefully, we eased Ted back up on the seat. He was okay. That was a close call. We should probably head back soon.

But as it happened, we were in an ideal little weedy niche, and so I asked Ted if he wouldn't mind if I cast a few times while

I was still standing in the water. "By God, no, I don't mind! Fish! Use my rod!" So I picked up his Helios and cast—incredibly smooth, the action of this rod—toward a small pool of open water bordered by dense plants that resembled miniature stalks of corn. A broad V shot toward my fly, and the water erupted, and I had a beautiful bass on. I love catching fish like this, in shallow water surrounded by vegetation—the challenges are considerable. I landed the bass with Ted watching and beaming. I felt like the subject of a fishing show on TV, with a live audience of one. This time, I didn't mind.

In truth, I also felt a little guilty, archetypally fishing while Ted recovered from his humiliating fall. But I couldn't resist, and there was another pool beckoning to me fifteen feet further down the dreamily lush shoreline. I dropped my fly right in the slot under an overhanging downed tree, and another bass burst out of the cover and took my fly. Another microadventure of coaxing the muscular fish through the weeds and to me. "Un autre!" Ted shouted. Or maybe it was "Un otro!" I couldn't quite tell, but it was definitely another fish.

I got back in the boat and rowed Ted around the far end of the lake, looking for old landmarks and underwater structure that he knew from years ago. I could tell Ted was done fishing, rattled after his plunge onto the floor, and so I tacitly angled the boat back toward the launch and headed that way. It had been a good morning—seamless, really, in so many ways. Our trip back was uneventful, and Ted was able to stand and make his way up the trail and to his truck without further incident. I pulled the boat back, and we loaded it up without trouble. Driving back, we recalled the highlights: the biggest fish we caught, and the ones we missed, which were no doubt even larger.

Ted thanked me for the trip, and I thanked him, and he scoffed and said we might have to start a mutual-admiration society. As I climbed out of the truck, Ted said, "Chris, I don't think I ought ever dare do that again." We shook hands, I got my rod and pack out of the boat, and I headed up the hill toward our house.

This story about Ted doesn't really go anywhere else, and I'm not sure if it has a clear lesson, or further meaning. But so much of fly-fishing is like this. Ninety percent of the time it's undramatic, and the end is always anticlimactic. More than this, I've had the chance to tarry with Ted, to feel fully alive, in this elastic time before death—alive, together, in the face of Ted's own self-acknowledged mortality.

Fly-fishing is an alternative kind of inhabitance. But once again, it's not exactly an escape. This is the world where we live—our one chance at life. As Octavia Butler voices through Lauren Olamina in *Parable of the Talents*, "What a cruel and terrible thing escape would be if escape were possible" (403).

Yet if you look at fly-fishing magazines and catalogs, it certainly *looks* like a romantic escape. If fishing *were* about escape, what kind of escape, escape from what? Toward the end of this book now, and the end of the summer during which I have been writing it, I am going out to fish one of my lakes again—the same lake you've read about many times in these pages. Escaping back out into the familiar.

I wake up at 5:30, a little past when I like to head out, but the sun is rising later these days—or rather, our planet is tilting on its revolution—and it will be fine. It is an atypically cool late

July morning—my car's computer tells me to watch out for icy roads!—but it will be in the eighties by midmorning.

There is a blanket of mist over the lake when I get to the place where I wade in, and it looks perfect. Even so, I am skeptical because this late in the summer the fish can turn off inexplicably, especially before scorching days. I walk out up to my waist and cast at the familiar reed lines, the walls of sedges, the jungles of pondweed and canopies of lily pads. I wade out deeper—I'm going to wade across the widest point of this bay today, a route I have yet to take. I should say here that my wading of these lakes has taken place over a dozen years, incrementally, feeling out the silty areas and the bottomless pits, backpedaling frequently and respecting the glacial shiftiness of this region as I've explored its inland bodies of water on foot.

I came out here today to think once more about why I go fly-fishing—what I'm escaping from, or toward, when I practice this ... sport? Activity? Pastime? Hobby? Obsession? Escape? I'm so far into this, and I don't even know what to call it.

Out on the water I begin to stitch things together in my head. Fly-fishing is not an escape from civilization or society. If anything, as in Henry David Thoreau's *Walden*, those things become amplified out here. The jet cruising thirty thousand feet above me; the joggers on the dirt road through the woods, who at first sounded like coyotes, until I realized that the yipping and cackling were coming not from wild canines but from exercising *Homo sapiens*; the gobs of fishing line that I get tripped up on as I wade, that I pull up from the depths and carefully (because of rusty hooks on the ends) wind into balls and put away in my vest for proper disposal, if there is such a thing; the thrum of charter fishing boats out on the big lake, hundreds of gas-powered horsepower churning the water ten miles away; my own car, parked off near the edge of the cove, its teeny red security light blinking comically on the dashboard; my phone buzzing in my chest pocket, alerting me to a text message of extreme minor urgency. No, human presence is all around here.

Is fly-fishing an escape from the self? Hardly. As Jan Zita Grover observes toward the opening of *Northern Waters*, while learning to fly-fish she makes a "quelling discovery, that I probably didn't like myself very much. I was impatient and almost unimaginably cruel to myself while learning the basics of streamcraft" (4). Grover's word *streamcraft* stands for learning to read moving water and merge with it using a fly rod—it's basically the art or skill of using the weighted line and a long, limber rod to propel a fly on a line to an exact spot. And yes, this practice can invite intense self-loathing. My own recourse is to sarcasm and swearing when my backcast hooks a standing reed tip or tree limb, or when my line doesn't turn over in the last instant and the fly plops inelegantly several feet away from where I'd been aiming. Managing a fly line puts the self under a severe microscope, and every action and thought becomes suspect, infected, saboteur.

Fly-fishing relies on a series of knots: backing to fly line, fly line to leader, leader to tippet, tippet to fly. Each point of connection demands a different knot. Each knot, a potential weakness. Tying knots—or worse, seeing them fail—can also open up the floodgates of self-criticism. John Merwin, in his Trailside Guide to fly-fishing, has an efficient section on knots, including a sidebar on "wind knots" that makes me laugh: wind knots are usually simple overhand knots that end up in the leader or tippet when the backcast pulls the accelerating fly through a loop in the line, and then tension (from a fish or just the retrieve) tightens it. But as Merwin notes, calling it a "wind" knot is deceiving: "They have nothing to do with the wind. They are caused by sloppy casting" (93). For some reason this line pierced me, and I hear it reverberating in my inner monologue as I fish and inevitably indulge in *sloppy casting*. I told Glen about this line, and now he chants it with me as we fish, whenever our casts fail: *Sloppy casting!* It's really a harsh analysis, because it's not as if we're *trying* to be messy. It's often a matter of standing chest-deep in incredible densities of vegetable matter while trying to cast to a perfect spot, and there are simply too many obstacles to execute a neat cast. And so often, it *is* the wind that makes it all worse! Here I am, making excuses for my sloppy casting—always inevitable at some point on any of my trips. I'm thinking about this now because I find myself in a tall stand of reeds, surrounded by fish and fishy spots, and plagued by these gorgeously green fibrous obstructions. My line is both the solution and the problem.

Maybe this is why I was drawn to Greg's poem and wanted to use it at the beginning of this book: "this ploy to sense another through a line." That's what I'm doing here! Both as I fish and as I write about fishing. Trying to connect with the fish that enchant me and attempting to connect with *you*, the reader, to communicate my strange, if predictable, even repetitive, obsession.

Fly-fishing is something that gets a weird reputation. It's the kind of activity that causes people to mythologize it; consider how the celebrity chef Anthony Bourdain describes one of his early role models in *Kitchen Confidential*, a "pasta man" named Dimitri, "a man of esoteric skills and appetites: a gambler, philosopher, gardener and fly fisherman" (46). It hangs there at the end, a loaded cipher. Then Bourdain returns to it, a page later: "As a fly-fisherman, Dimitri made his own lures; this obsessive eye for detail carried over to his food" (47). The chef's expert qualities are supposed to be translated through his talent for tying flies—a translation of inscrutable skills.

The obsession doesn't conclude with the fish at the end of my line, nor can it be easily translated to another more prominent outlet or professional topos. It's messier. John Law and Marianne Lien's essay "Denaturalizing Nature" explains fly-fishing as both "a form of Romantic subjectivity [and] also a set of embodied skills": "Contest, tug-of-war, fight, game of deception, feint, and counterfeint—these are some of the images that characterize the joy of fly-fishing" (136–37). But for Law and Lien, fly-fishing ends up leading to a complex Nature-Culture entanglement.

I experience this entanglement daily; it decenters and disorients me. Fly-fishing includes the deerflies strafing my hands and head, the fly-catching redwing blackbirds and kingbirds, the darting kingfishers, the great blue heron fishing twenty feet away from me (it reminds me of our old Siamese cat Henry, the way it creeps and stalks), the sandhill cranes chortling across the lake, the pair of loons low-swimming nearby, the coffee-table-size snapping turtles that come to check me out, the water snakes that slither by, all the trees surrounding the lake, a particular white pine on a point whose top rakes at an angle, like a giant's bonsai, the pungent wafts of sulfur that bubble up from beneath my footsteps, millennia of geological layering released in gaseous ether . . . All these things form my obsession, my fly-fishing practice. The escape is into this multifaceted familiar place, focused around the fishes but opening up and out to a kaleidoscopic wilderness that absorbs and refracts me, my humanity—and humanity at large. The fly line exposes the sloppiness and connects us all. And these lines here, Socrates's nightmare, record my reflections on water and in prose.

I cast toward a clump of reeds that juts out away from the line. As my blue damselfly imitation lands, it is immediately engulfed—a strong bass pulls toward the weed cover, and I ease it back to me, stripping line and using the bend of the rod to coax it within arm's reach. As I struggle to bring the bass in, I see two others following right beside it. I love when they do this: the bass racing together, protective or competitive or a bit of both, or something altogether different. I gently cradle the fish's belly in my palm in the water and pluck the hook from its jaw. The other two bass hold down by my feet, waiting for their mate. These bass are so slender and pretty that I think of them as trout-like bass. I release the caught bass, and it swims down to its friends; the three fish swim lazily together back into the cover, a collective that I feel intimately a part of even in my eccentric adversarial role above the surface. This is my awkward condition, a provisional escape.

I want to end this book with a formula for a different kind of fly-fishing. Actually, it's a subset of the type of fishing I've been writing about this whole time, but here I am going to boil it down. Because fishing has become complicated. Species populations on the brink of collapse, microplastics swirling in vast gyres, prodigious oil spills, a global pandemic, ice shelves melting, unacknowledged structural racism, displaced peoples, ever more hurricanes . . .

In these gloomy times, it's not always realistic, much less affordable or even ethical, to go destination fishing for rare fish species. Setting records is passé; now we know that this is simply what capitalism *does*, everywhere and always. So instead of these kinds of fishing, I find it's better to set out through your backyard or down the street and go after what's nearby. I call it *small-fishing*. It's the opposite of trophy fishing, the obverse of long casting in remote locations. Yet small-fishing has its own treasures in store, and it's addictive beyond your wildest dreams. It's also a version of what Thomas McGuane advocates pithily in *The Longest Silence*: "A case could be made that fishing home water and taking good care of it is what an angler should do" (277).

I first did this outside of Bozeman. While others were out fishing for big rainbow trout on the Madison out in the valley, I would bike a couple miles out of town to the East Gallatin and coax small brown trout out of submerged car windows in the garishly reinforced banks of the creek. Or sometimes Greg and I would pack our lightest rods and drive up Hyalite Canyon, fishing little pools for minuscule brook trout: Greg called them "cookies." Once one of Greg's colleagues asked us to stock an ornamental pond on his property, and so we carefully caught a couple dozen tiny brookies and brought them sloshing in a cooler up a meandering drive to a pristine alpine yard, tiptoed across the lawn, and dumped the fish into the sculpted recirculating body of water. I slid precariously as the perfect bank collapsed beneath me. I massaged the sod back in place before we left. Now Greg's colleague could small-fish—if he fished, which he didn't, but at least he reported back that he would often enjoy sipping his morning coffee while watching his private trout rising to insects on the pond's surface.

The goal is to target the smallest fish you can find, close to where you live. You're looking for bodies of water that are sometimes barely more than puddles but that are teeming with small fish. Garish beaver channels, urban canals, seasonal inlets, minor bogues, overlooked roadside ponds, airport-runway spill-off rivulets—these sorts of waters. James Barilla concludes his picaresque book *West with the Rise* self-deprecatingly, admitting that in the end he's a mere "smolt fisherman"—*smolt* being the smallest trout. I applaud this and want to embrace it as an actual method.

Head into a nearby forest or walk down the street, tracing the curvature of the land, occasionally traipsing over concrete. Follow it to a water source—one will appear eventually. Sometimes you'll get disoriented or lost, or feel foolish wandering down a derelict alley with a fly rod, but that's part of the adventure.

One time I took Julien out for a midday fishing trip on one of our small lakes up in Michigan, an ill-conceived idea as it was sunny and a little too windy. But we had the canoe on our car, and he was excited to go (for once), and so we went. At first he was upset that his big K&E rubber worms weren't working; we only had one shoreline we could fish reasonably, what with the chop on the lake. I was at risk of failing, yet again. But soon enough we discovered that vast schools of three-inch bass and bluegills were holding by the sedges twenty feet out from shore. We would see them startle from time to time as a larger fish tore through them or as a kingfisher swooped down from a nearby branch. So we changed our tactics, and I plucked out some diminutive EP minnows I had tied back in New Orleans, for use in Bayou St. John—miniatures of the larger minnows I tied for big bass and pike, no more than two centimeters long, on size 16 hooks. I attached a casting bubble to Julien's line— Greg taught me how to do this, back in Montana on Hyalite Lake, fishing Pink Lady flies for cutthroats. Julien and I cast toward the frantic ripples and then would rapidly pull the flies through the schools of younglings, creating the sensation of out-of-place prey. The small fish would race after our flies and compete to eat them; it was an object lesson in seeing these fishes' behavior at an early age: doing exactly what full-size bass and bluegill do, at a smaller scale. Julien was ecstatic as we caught one after another of these dainty fish. And they were all hooked cleanly and released quickly. Small-fishing saved me from another botched session.

For small-fishing I often use a diminutive seven-foot tenkara rod made in Japan that I found online for $5.99. I thought the price *must* have been a typo, but I ordered it on a whim; it arrived, improbably, in a neat parcel four weeks later—a palimpsest of international customs stamps around my address. I attached a remnant of an old floating line to the lillian, for casting, adding a ten-pound test leader and then four-pound for tippet.

Old ravaged flies work just fine. A scuffed-up tiny popper or pockmarked ant, or a basic shredded caddis. For waters with some depth, thread a simple brass bead onto a small hook, then tie a single deer hair or peacock herl onto the shank and wrap it with whatever flashy remnants you have lying around—and you're done. It doesn't have to look like much to do the job.

These fish have never seen a fly before, and they are opportunistic feeders.

A favorite place where I small-fish is a spring-fed cedar swamp that follows a labyrinthine path before emptying into Lake Michigan, near my childhood home. I cross a meadow and trace a network of deer trails that skirt the aquatic zones behind the dunes. The pools of water occasionally widen and deepen so that they become liminal habitats for frogs, turtles, leeches, redwing blackbirds, ducks—and small fish. Chubs, mud minnows, shiners, dwarf sunfish, and rogue perch inhabit these structure-choked waters.

I'm sight-fishing for schools that cluster around downed trees. There is no gorgeous backcasting here, only little wrist flicks and quarter roll casts, so as not to hang up on the matrix of dead cedar limbs and overhanging oak branches. I'm basically competing with the kingfishers and green herons.

Today the whole surface is dappled with feeding, and fish strike my tiny popper each time it alights. I hook a nice three-inch chub, and it bulldogs for the bottom—ounce for ounce, chubs must be some of the best fighters out there.

In a deeper cove blooming with milfoil, I see a mysterious giant carp roving among some silver-dollar bluegills, like an awkward cousin visiting from out of town. My heart races—but I am there for the small fish. I flick a tiny nymph toward a snarl of logs to coax small perch from the darkness where I know they gather. There is a flash, and suddenly I have a thrashing four-incher on, bending the delicate rod in a deep arc.

Now I'm stalking an elusive school of hybrid sunfish—mostly pumpkinseed but something else, too. They compete over my beadhead, attacking it with gusto, but their mouths are so petite they struggle to devour the whole bug. I finally hook one, a two-inch plump beauty. I carefully remove the tiny hook from its sapphire jaw and set it back into the pond, where it darts off toward the beaver dam.

A fifteen-minute walk through the woods and then across the meadow, and I'm home. I carefully remove the ticks that have come along for the ride, for another kind of fast food—I am their Happy Meal. Tonight I'll dream of chubs rising to minuscule spiders, and of sunfish chasing alien nymphs. As soon as I can, I'll be itching to get back there again, back to the nearby, to fly-fish for the smallest.

One more time. One last trip out to my lake before we pack the car and head back to New Orleans. "My" lake—not remotely. This lake existed for several thousand years or more before I was born, and will long outlive me, until dune succession buries it beneath beech and maple trees. But for this minor duration, I will continue to visit it and fly-fish its shorelines. Today I didn't catch a single fish. Getting "skunked," it's called—and oddly, there was the aroma of a skunk as I drove to the lake through the predawn mist early this morning. An olfactory omen, overdetermined.

I fished a particular line of pondweed stands that I know well—but not today. Yesterday a torrential rain dumped on the peninsula, causing flash flooding and raising the lake levels by six inches. All the familiar markers were invisible; it was a different lake. The water temperature had shifted, too, undoubtedly. I didn't see a single bass or sunfish. I saw pike slashing at schools of shiners, belly-flopping and causing a wild commotion in the swampy corner of the lake. But the pike seemed to be gorging on the minnows exclusively, and I could barely get them interested in any of my flies. I threw everything I'd brought—minnows, deceivers, poppers, frogs, sliders, divers, dragonflies—and had only a few fish swirl behind my fly as I retrieved it before tearing off and leaving sink-size boils in the water before me. A kingfisher repeatedly plunged into the water behind me, breakfasting. As I turned and cast toward a slot between sedges, my green frog-pattern fly hit the water only to be dive-bombed by the kingfisher—alarming me and causing me to quickly yank the fly out of the bird's attack vector. That was the only "bite" I got today: a vertiginous strike indeed. I had only my heavier eight-weight rod, and so I couldn't even catch small fish. But I didn't even see the schools of little three-month-old bass today.

The swollen lake sabotaged my fishing, but that's okay. I'm still processing the sudden storm: our rutted-out driveway, a nearby river having exceeded its banks, a county road blocked by a firetruck, the roadside creek bulging over the underpass, and an inundated playground in the small town five miles away. I was explaining climate change to Julien, how it affects different people and regions disproportionately and with uneven effects—this is why some people don't believe in it and why others are working swiftly to adjust their lifestyles around it. Did climate change trigger this mini-monsoon? Quite probably, though it could be all but impossible to draw a straightforward causal chain of events to explain how. Climate change is so big, and the results so scattered and seemingly isolated—even when they are devastating. The results of climate change are as dispersed as the driving forces of it are interconnected: the acceleration of fossil fuel burning and the rampant growth of global capitalism. We talked about going back to New Orleans, with hurricane season in full force. We talked about our friends Jon and Brooke and Emmett and Elsa, who live in Oregon under the billowing smoke haze that had wafted three thousand miles to us on the jet stream just a couple weeks ago. We talk about Greg in Bozeman, who tells me that the smoke is so bad they have been advised to stay inside there—he ends an email to me with "Seems like we're always sheltering in place." The Delta variant of the coronavirus is causing new infections to surge. Julien wondered, "What's COVID have to do with climate change?" Well . . . I mumbled an incoherent answer.

All this was on my mind as I fished in futility this morning. I failed utterly, and this failure hangs above me like the setting yellow full moon over the lake as the sun comes up through the pines behind me.

Fly-fishing is challenging, and all the challenges of the world come along with it. There is no escape, even if this practice does open up sublime portals to other forms of coexistence, entanglement, and inhabitance. Fly-fishing has made a fool out of me today, standing out in the shadowy water up to my chest, unable to do what I thought I knew how to do . . . feeling the ground give way beneath my boots. Two great blue herons squabbling down the shoreline suddenly lift off and angle toward me, flapping within a few feet of my head so that I duck instinctively. They land in the shallows right behind me, continue to bicker for a minute, and then soar away, taking their scrap back to the lily pads in the far corner of the lake.

AFTERWORD

MINOR-FISHING LESSONS

I'M WATCHING WATER DIMPLED with concentric circles from where bluegills look to be rising. Those disturbances might not be bluegills, though; they might be the backs of gar, gulping oxygen. Or they might be pipefish feeding on tiny crustaceans. Or schools of mullet slurping up microplankton. Or nothing at all.

Either way, I am pulling handfuls of line off my reel, preparing to make my first cast. The line gets easily caught up in the clovers on the ground, and I'm trying to keep track of it as I look out over the surface. I have crept up to the shore and am deciding whether to cast parallel to the bank toward the "watch for alligators" sign or across the end of the bayou, toward the square cement top of an inlet pipe. The sun is coming up on the abandoned hospital across the street, a huge graffitied PLEASE illuminated across the brick roofline. I am fly-fishing in this most unlikely of places.

It takes me six minutes to walk to Bayou St. John, near my home in Mid-City New Orleans.

I walk there to fish the end of the bayou, a cement-lined channel about thirty feet across, greenish water and occasional birdlife in residence: mangy blue herons, a solitary sandpiper, a roving pelican, sometimes an osprey. Among these creatures I scout for small fish in the murky shallows, casting surface bugs and darting streamers when I see signs of fish. I'm small-fishing, or as my friend Margret renamed it when I described it to her, *minor*-fishing. The whole practice is at once heightened and deemphasized, focused and decentered. Coming off my Michigan summer, though, and teaching through an ongoing pandemic, this daily hour-long pursuit has kept me sane.

Here's what I've learned this fall, minor-fishing here:

I often mistake things I see for other things. I see a circular sandy spawning bed on the bottom; only, when I get closer, I see it's a sunken white trash bag, faint red letters undulating *THANK YOU*. I see a dead fish in the water, silvery and still; only, it's really a candy-bar wrapper. I see a turtle's head in the water, looking at me; only, it's the moss-covered toe of a high-heeled shoe floating. Things are not what they seem.

When I fish in Michigan, I look for structure: reeds, submerged logs, lily pads. Here I find myself targeting the floating forms of face masks, pill bottles, and deflated soccer balls. I cast toward murky shapes in four feet of water, and the edge of the sloped cement embankment, where fish hang out. Adjust to your environs.

I once felt a pull on my line, and when I set the hook, a blue surgical glove came flying out of the water toward me. It flapped and waved, almost in slow motion, as it approached me. I dodged it, just barely. Another time, I caught a half-submerged empty Doritos bag, Spicy Nacho flavor, adorned with an elaborate ad for a video-game: CALL OF DUTY VANGUARD | WARZONE | UNLOCK DUAL 2XP WEAPON WITH EVERY BAG. The background of the chip bag was a topographical map of a virtual world. And here I thought I was just fishing.

One time I cast a small blue popper toward a tangle of branches in three feet of water, and as my fly hit the surface, I noticed it had landed six inches from an actual blue damselfly. A fish raced up right at that moment and gulped the true damsel down; then the fish came back for my doppelganger fly. I felt its pull but didn't manage to set the hook. Yet the repetitions were enough. The magic had worked.

Cypress-tree branches floating in the bayou can bear an uncanny resemblance to alligator heads. And it is a real possibility, here, for alligators to appear. After Hurricane Ida, someone was eaten by an alligator that turned up in their yard. Driving back into town after our week of evacuation and refuge in St. Louis, I saw an alligator squashed terribly on the side of the highway, making its own return after the storm. Another time while fishing I saw a large lizard on the cement slab and got out my phone to take its picture, only to realize once I got within a few inches of it that, no, it was a baby alligator. And its mother was probably nearby. We live in an uneasy truce with these creatures here on the dirty coast.

In the bayou I have caught redear sunfish, cichlids, bluegills, yellow bass, and largemouth bass. I see schools of mullet cruising a few inches beneath the surface, but they are almost uncatchable, sifting feeders that they are—except with a fly that I am told resembles a wad of bread dough, if dropped directly in front of the mullet as they feed. I have also seen a small pipefish, and several giant gar. I watched the taciturn fisherman who never speaks to me catch a redfish. A few days later, when I waved at him, he nodded at me. All in this small finger of brackish water connected to the Gulf of Mexico.

Mind the backcast! This isn't like casting while wading in a lake in the middle of relative nowhere, or on a wide river. People are jogging, doing yoga on the grass, playing volleyball, just sitting smoking or drinking boxes of wine at 8:30 in the morning, throwing discs to leaping dogs, walking and talking . . . I have learned to be aware of other people's practices as I make my way around the end of the bayou.

The bridge across Orleans Avenue has fish lurking beneath it; they cluster near the concrete pilings holding up the expanse. But they are hard to reach, with the obstacles around this intersection. Last time I tried, I inadvertently snagged a traffic light behind me, pulsing red. STOP.

Don't fish after they mow. Once a week an efficient grounds crew hacks down the scraggly grass and weeds bordering the bayou, and for the next day there will be a scrim of clippings over the water. This could be good for fishing, but with the trimmings come along insects and other edibles, on which the fish gorge. Or even if not, my flies get all hung up in the detritus and lose their appeal. Whatever the case, minor-fishing is always impossible after they mow. Or rather, it becomes even more minor—ineffective, a little charade.

I've been snapping quick photographs of the fish I catch, setting their bodies against the handle of my rod, with my small three-weight reel at the bottom for scale, all lying against the impossibly green clovers or the cement conglomerate. But two times when I've shown people pictures of my Bayou St. John fish, they ask what the sheriff badge is for. I was confused by this until I realized that my Lamson reel's spool resembles a star—it looks like an old western motif. My pictures, meant to show off my catches, have caught me in an absurd story beyond my making.

BIBLIOGRAPHY

Barilla, James. *West with the Rise: Fly-Fishing across America*. Charlottesville: University of Virginia Press, 2006.

Bourdain, Anthony. *Kitchen Confidential: Adventures in the Culinary Underbelly*. New York: Ecco, 2007.

Butler, Octavia. *Bloodchild and Other Stories*. New York: Seven Stories, 2005.

Butler, Octavia. *Parable of the Talents*. New York: Grand Central, 2019.

Cather, Willa. *Death Comes for the Archbishop*. New York: Vintage, 1990. First published in 1927 by Alfred A. Knopf.

Cohen, Jeffrey Jerome, and Lowell Duckert, eds. *Veer Ecology: A Companion for Environmental Thinking*. Minneapolis: University of Minnesota Press, 2017.

Darkes, Jerry. *Fly Fishing the Inland Oceans: An Angler's Guide to Finding and Catching Fish in the Great Lakes*. Mechanicsburg, PA: Stackpole Books, 2013.

De la Valdène, Guy. *On the Water: A Fishing Memoir*. Guilford, CT: Lyons, 2015.

Gamerman, Amy. "How the Rich Fish." *Wall Street Journal*, June 1, 2017.

Grover, Jan Zita. *Northern Waters*. St. Paul, MN: Graywolf, 1999.

Haraway, Donna. *Staying with the Trouble: Making Kin in the Chthulucene*. Durham, NC: Duke University Press, 2016.

Keeler, Greg. *Trash Fish: A Life*. Berkeley, CA: Counterpoint, 2008.

Keeler, Greg. *Waltzing with the Captain: Remembering Richard Brautigan*. Boise, ID: Limberlost, 2004.

Kurlansky, Mark. *The Unreasonable Virtue of Fly Fishing*. New York: Bloomsbury, 2021.

Law, John, and Marianne Lien. "Denaturalizing Nature." In *A World of Many Worlds*, edited by Marisol de la Cadena and Mario Blaser, 131–71. Durham, NC: Duke University Press, 2018.

Le Guin, Ursula K. *The Beginning Place*. New York: Harper and Row, 1980.

Maclean, Norman. *A River Runs through It and Other Stories*. Chicago: University of Chicago Press, 1976.

McGuane, Thomas. *The Longest Silence: A Life in Fishing*. New York: Vintage, 2019.

Merwin, John. *Fly Fishing: A Trailside Guide*. New York: Norton, 1996.

Miller, Lulu. *Why Fish Don't Exist: A Story of Loss, Love, and the Hidden Order of Life*. New York: Simon and Schuster, 2020.

Natural Area Guide. Leland, MI: Leelanau Conservancy, 2015.

Odell, Jenny. *How to Do Nothing: Resisting the Attention Economy*. Brooklyn, NY: Melville House, 2019.

Schaberg, Christopher. *The Textual Life of Airports: Reading the Culture of Flight*. New York: Bloomsbury, 2013.

Singh, Julietta. *Unthinking Mastery: Dehumanism and Decolonial Entanglements*. Durham, NC: Duke University Press, 2018.

Thoreau, Henry David. *Walden*. Boston: Ticknor and Fields, 1854.